Environmental SCIENCE

How the World Works and Your Place in It

Jane L. Person

Second Edition

Study Guide

J.M. LeBel Enterprises
Dallas New York Toronto Edmonton

Environmental Science
Study Guide

Components:

- Textbook
- Teacher's Manual
- Issues and Investigations
- Study Guide
- Answers to the Study Guide
- Test Program

ISBN 0-920008-74-7
Study Guide

J.M. LeBel Enterprises
6420 Meadowcreek Drive, Dallas, Texas 75240
10335-61 Avenue, Edmonton, Alberta T6H 1K9

CONTENTS

To the Student

This study guide should be used to improve your reading comprehension and as a review guide for tests. The manner in which you use the study guide is important to your success. By using the following process, your understanding of the concepts presented will be improved, and your test scores will be higher.

1. When a section or case study is assigned, survey the entire assignment and read the headings.

2. Return to the beginning of the assignment and read the introduction or the first segment of the assignment.

3. Answer the questions in the study guide that correspond to the segment just read.

   The questions in the study guide are arranged in sgements that correspond to the segments in the textbook. Page numbers are given to help you locate the information quickly.

   Bold print will help you identify possible answers or choices you must make. Underline or circle the correct answer or write the letters or numbers in the appropriaite spaces.

   Use complete sentences when writing answers to short answer questions.

   When asked to state your opinion on an issue, remember that opinions are not to be judged as right or wrong answers. They reflect your feelings or beliefs about an issue. Be sure to read the information so that you acquire as many facts as possible before forming your opinion. Your opinion may change as you aquire more knowledge about an issue.

4. During class or group discussion, be sure to correct any answers that are incorrect and answer any questions that you could not answer before discussion.

5. When the answers in the study guide are complete, use it to review for tests and as a resource for other assignments that your teacher may give you.

   Have a good time learning about your World and How It Works!

Name ______________________ Date ________ Class Period ____

Unit I — 1.1 Ecology — a Study of Relationships

Introduction [3]

1. **Ecology** is the study of ________________ that are living in their ____________ ________________.

2. How does an ecologist's work differ from the work of other biologists?

____

____

____

3. An **ecologist** might also be called a ________________ biologist or a ________________ biologist.

4. Check the statement that **best** describes what an ecologist studies.
   [ ] An ecologist studies living organisms.
   [ ] An ecologist studies the relationships that exist in an ecosystem
   [ ] An ecologist studies the environment.

5. The **biosphere** refers to the layer of ________, __________ and __________ where life exists.

6. Describe two ways that scientists collect information (data) that tells us about the health of planet Earth's biosphere.

____

____

Viewing Planet Earth from Space [3]

7. What information would you present if asked to justify the expense of launching a satellite that would gather data from planet Earth?

____

____

____

8. What if scientists wanted to know answers to these questions: "How fast are the world's tropical rain forests disappearing?" and "How much habitat remains that can support populations of grizzly bears?" Which of the methods described in question 5 would be the best source of the information that the scientists need?

____

____

9. Identify the technology represented by each of the following abbreviations and describe how scientists use each technology to study the health of planet Earth.

EMR ______________________________________

____

____

TOMS _____________________________________

____

____

Name ______________________________ Date __________ Class Period ____

Biomes—Climatic Zones of Life [4]

10. If the biosphere is a puzzle, the pieces of the puzzle represent the ______________. Each **biome** has a distinctive type of native ______________ life that is determined by the ______________ and ____________ type.

11. The **climate** refers to the "average" ________________ of a biome. It describes the ______________ and ________________ of precipitation and the normal range of ________________________.

12. **Think about it!** What are the features of climate that affect the biome's
    soil type? ____
    plant life? ____

13. Using the information from the chart, identify the biome described.

____________________ This biome has hot and cold seasons, periodic droughts, and very deep topsoil that is rich in nutrients.

____________________ The soil and plants are similar to the biome described above, but the amount of rainfall is greater and there is no cold season.

____________________ We tend to think of this biome as being hot, but it may be very cold. Plants must be adapted to store moisture and withstand drying winds.

____________________ Most rain falls during the short summers in this biome. Plants must tolerate the nutrient poor, acidic soils. Although the soil freezes in winter, it is not permanently frozen.

____________________ Conditions in this biome are even more severe than the biome described above. Only the surface of the soil thaws during the summer.

____________________ Seasons in this biome are determined by the amount of rainfall rather than temperature changes. Plants grow throughout the year. The soil has few nutrients and becomes hard if exposed to the sun.

____________________ The biome where you live.

Ecosystem—a Functional Unit [4]

14. An **ecosystem** is a group of ________________________ that ______________ with each other and with their ________________________.

Name ______________________________ Date ________ Class Period ____

15. List three ecosystems near your home and three ecosystems that you would like to visit.

Local Ecosystems	Ecosystems I would like to visit.
1.____________	1.____________
2. ____________	2. ____________
3. ____________	3. ____________

16. Natural ecosystems are self-sustaining and do not need any input from humans, but they must have a supply of ____________ and __________ (usually in the form of sunlight).

17. **Think about it!** Lightning often associated with thunderstorms is the cause of many wildfires. Although fires are destructive, they are essential to the prairie ecosystems. Explain the connection.

____
____
____

18. Compare a natural ecosystem to a human-made ecosystem.

____
____
____

Habitat—a Place to Call Home [6]

19. An ecosystem consists of both living and nonliving parts. List the living and nonliving parts of a goldfish's ecosystem.

Insert picture of goldfish bowl from page 1 of study guide here.

Living	Nonliving
____________	____________
____________	____________
____________	____________
____________	____________
____________	____________

20. "Nothing can live alone." Beginning with the word "plants" draw a simple concept map that shows the connections between plants and other living and nonliving parts of an ecosystem.

PLANTS

Name ______________________________ Date __________ Class Period ____

21. The major food in the squirrel's diet is the supply of nuts produced by mature trees in a mixed oak-hickory forest. Identify three physical factors that might limit the squirrel's population.
(1)____________________ (2) ________________ (3) ________________

22. The special place within an ecosystem that an animal calls "home" is called the animal's ______________.

A potential "home" site might be rated as "good," "fair," or "poor." Identify four needs that must be met for a habitat to be classified as a "good" habitat for a species.
(1)________________________ (2) ________________________
(3) ________________________ (4) ________________________

23. **Cover** refers to places that provide animals with protection from ______________ or ____________________________.

Populations of Woodpeckers [7]

24. If organisms are members of the same **species** they can ____________________ ____________________________. Members of different species normally do not breed because of differences in ________________ or __________________.

25. Because they do not breed, downy woodpeckers and hairy woodpeckers are classified as different ______________ of woodpeckers. The total number of downy woodpeckers in a forest is the ______________________ of the species.

26. When counting woodpeckers in a forest, is it appropriate to add the number of downy woodpeckers and the number of hairy woodpeckers? ________ Explain.
____
____
____

27. Describe the preferred habitat of these woodpeckers by completing the following chart.

Species	Location of Habitat	Type of Tree
Downy woodpecker		
Hairy woodpecker		

Do the downy and hairy woodpeckers compete for habitat? _______

Name ______________________________ Date __________ Class Period ____

More Cavities Needed [7]

28. Describe how each of the following changes affected the quality of the bluebird's habitat and the impact on the bluebird populations.

Human Activities	How activity changed the bluebird's habitat	Probable Change in Population
Early settlers cleared forests to create fields		
Use of insecticides		
Fences built using wooden posts		
Steel posts replaced wooden fence posts		
Invention of chain saw		
Cattle population increased.		
Construction of roads and developments		
Starlings and sparrows introduced to U.S.		
Nesting boxes placed next of open areas		

29. Identify two concerns that must be addressed when providing nesting boxes for bluebirds to decrease competition from other species.

(1)____________________ (2) ____________________ (3) ____________________

30. Compare the habitat requirement of the bluebird to the habitat requirement of the wood duck.

____
____
____

Finding the Right Neighborhood [9]

31. The activities of early settlers improved the habitat for the bluebird population by creating clearings in the mature forests. Identify three other species that also benefited when early settlers cleared the forests and killed predators.
(1)____________________ (2) ____________________ (3) ____________________

32. Today clearings in the mature forests are created by the construction of roads and housing developments, but the presence of humans is causing a decline in some populations including the ______________________

33. Although the water in most streams in the Pocono Mountains is clean enough to support good populations of trout, the population of this shy aquatic mammal, the ________________________, is declining in areas where there is too much human activity.

Name ____ Date __________ Class Period ____

Living on the Edge [9]

34 Farming activities improve the habitat for the woodchuck or groundhog by improving the __________________________.

35. The ______________ or **edge** is the place where two ___________________ meet. Give an example of an ecotone. ____
____

36. There are **(more/fewer)** species and **(larger/smaller)** populations of each species at the ecotone or edge. This is known as the edge ________________ .

37. Human activities, such as construction of highways, shopping centers and housing developments often **(increase/reduce)** the ecotones or _________________.

38 When a farmer removes a fence row to enlarge a field, which species will be least affected—groundhogs, rabbits, or quail? __________________________ Explain.
____
____
____

39. The most critical threat to wildlife today is _______________________________.

Limits to Population Growth [10]

40. The **carrying capacity** refers to the ________________ of the population that an _______________________ can support at any given time.

41. A **limiting factor** is anything that prevents an _______________ in the size of the population. Thus limiting factors reduce an ecosystem's _____________________
______________________.

42. Identify at least one major limiting factor for each of the following populations.

Population	Limiting Factor(s)
Breeding pairs of birds	
Lemmings	
Lynx	
Snail Kite	
Spined soldier bug (good bug)	
Deciduous trees (maples)	
Rabbits	
Mosquitoes	
Loons	
Bobcats	
Oak trees in a mature forest	

Name ______________________________ **Date** ________ **Class Period** ____

Population Explosions [12]

43. Population explosions occur when there are no ______________________.

44. Check the phrase that describes the events that occurred after the release of the rabbits. When Mr. Austin released rabbits in Australia, the population of rabbits:
[ ] was quickly eliminated by predators.
[ ] increased until they destroyed their habitat.

45. The ______________________ eventually became the limiting factor for the rabbit population.

46. What caused the explosion of the mule deer population on the Kaibab Plateau?

____

47. The mule deer population increased until the population density was one deer per 7 acres. What effect did this have on the carrying capacity of the Kaibab Plateau?

____

48. Government officials established a ________________ program that reduced the population density of the mule deer to __________ per _________ acres.

Why did the biologists decide to lower the population density to this level?

____

____

Name ____ Date __________ Class Period ____

Unit I — 2.1 Food Chains and Food Webs

Introduction [15]

1. Plants can not make **(energy/food)**, but they can make their own **(energy/food)** in a process called __________________________. The energy for this process comes from the __________.

2. Organisms that can use the sun's energy to change non-food molecules (water and carbon dioxide) into food molecules (sugar) are **(consumers/producers)**.

3. Producers contain __________________________ that traps the sun's energy. Two types of producers are ______________________ and ______________.

4. Even though people "make" things, people are not producers. What is the important difference between a green plant "making" food and you "making" a cake?
____
____

5. We sometimes refer to vegetables and fruits as produce. A farmer takes "produce" to the market. You buy tomatoes at a "produce" stand or in the "produce asile" of the grocery store. Why do you think that vegetables and fruit are called "produce?"
____
____

6. Why are producers considered to be the foundation of nearly all ecosystems?
____
____

7. Organisms that lack the ability to change non-food molecules into food molecules are classified as **(consumers/producers)**.

8. Match the correct term with the statements below.

(1) carnivores	**(4) omnivores**
(2) decomposers	**(5) producers**
(3) herbivores	**(6) all of these**

____ A. organisms that can make their own food.
____ B. organisms that feed on animals instead of plants.
____ C. organisms that eat both plants and animals.
____ D. organisms that eat only plants.
____ E. organisms that break down and absorb dead organisms or their wastes.
____ F. organisms that are the basis of all food chains.

9. A food chain shows the flow of ______________ in an ecosystem.

10. The position of an organism in the food chain describes its ______________ or job in the ecosystem. It can be compared to a person's __________________.

11. What is the niche of each of the following organisms?

koala bear	____________________	brown bear	____________________
coyote	____________________	fungi	____________________

Name ________________________________ Date __________ Class Period ____

12. Match the correct term with the statements below.

(1)	**decomposer**	(4)	**predator**
(2)	**host**	(5)	**prey**
(3)	**parasite**	(6)	**scavenger**

____ A. organisms that hunt and kill other organisms for food.
____ B. organisms that supply food for parasites.
____ C. organisms that eat animals which have already been killed.
____ D. organisms that break down dead organisms and produce humus.
____ E. organisms that are killed to provide food for other organisms.
____ F. organisms that feed on other organisms while they are still living.

13. What is the niche of each of the following organisms?

mosquito	________________	house fly	________________
vulture	________________	grass	________________
algae	________________	bacteria	________________

Building a Food Chain [16]

14. The first step in any food chain is always a ____________________. The second step is always a ____________________.

15. The **primary consumer** is always a ____________________ and the **secondary consumer** is always a ____________________.

16. In a food chain or food web an --------> is read as ____________________

17. In one or two sentences describe the food chain shown in the diagram above.

____

____

18. Identify the organism that fills each of the following **niches** in the food chain described in question 17.

predator	________________	secondary consumer	________________
producer	________________	primary consumer	________________
herbivore	________________	carnivore	________________

Name ____________________ Date __________ Class Period ____

19. Could the flowers in the picture be replaced with a picture of an Indian Pipe? _____
Why or why not? ____
____

20. What is the niche of each of the following organisms?

mosquito	______________	house fly	______________
vulture	______________	grass	______________
algae	______________	bacteria	______________

21. A **(food chain/food web)** best shows the relationships that exist in an ecosystem. Ecosystems that are stable have **(complex/simple)** food webs.

22. Give the term which properly describes the niche of each organism as it is shown in the food web.

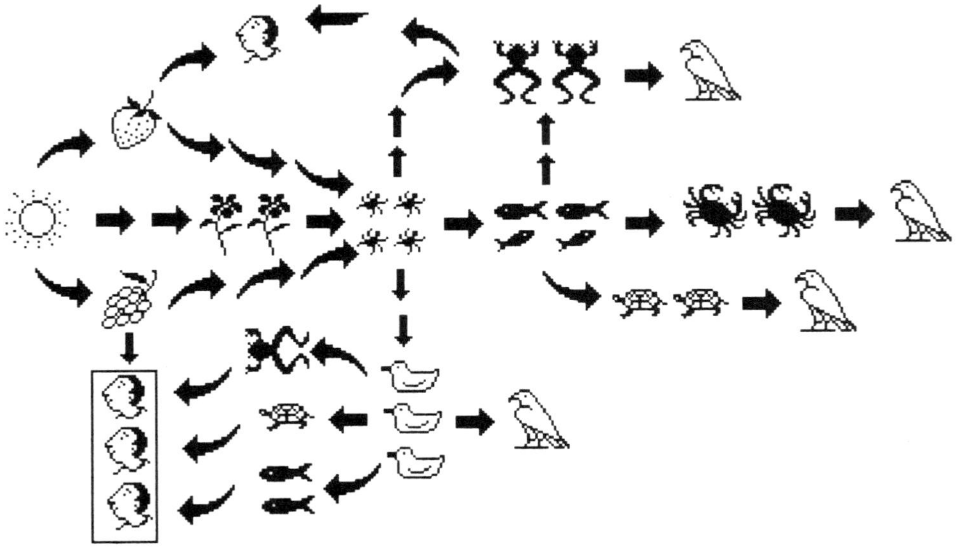

turtle	______________	frog	______________
hawk	______________	flower	______________
crab	______________	fish	______________
grapes	______________	insect	______________
human	______________	strawberry	______________

23. Is this a stable ecosystem? _________
Why or why not?
____
____

Name ______________________________ Date __________ Class Period ____

Unit I — 2.2

Spare that Snag!

Introduction [19]

1. A dead tree that is still standing is called a ______________.

2. Give two reasons why snags are becoming rare in some forests.

   ____

   ____

3. Provide evidence that supports the statement: "A standing dead tree is worth a lot to wildlife."

   ____

   ____

4. Most cavity nesting birds, including woodpeckers, eat ______________. The term that describes their niche is ______________.

5. Match each woodpecker with the proper description of its habitat.

   (1) common flicker (2) ivory-billed (3) pileated (4) red-cockaded

   ____ A. requires large tracts of old growth forest.
   ____ B. makes cavities in living trees in open pine forests in Southeastern U.S.
   ____ C. may nest in utility poles or abandoned animal burrows.
   ____ D. makes its home in smaller tracts of maturing forests.

6. Match each woodpecker listed in question 4 with the proper description.

   ____ A. This species is endangered and probably extinct.
   ____ B. This species is an indicator species.
   ____ A. This species is the most adaptable and therefore the most common.
   ____ D. This species is threatened by the conversion of forests to pine plantations.

7. During a hike through a forest you spot a pair of pileated woodpeckers. What does this tell us about the forest?

   ____

   ____

8. Complete the food chain below.

   ______________ -----------> ______________ -----------> **Woodpeckers**

The U.S.F.S. Policy [20]

9. Biologists working for the U.S.F.S. recommend leaving ________ snags per acre.

10. Trees are classified as **cull trees** if they ______________________.

    Why are cull trees important? ______________________________.

Name ______________________________ Date __________ Class Period ____

11. Why are forests with snags healthier than forest where all dead and dying trees are removed?

____

____

Life in the Old Hemlock [20]

12. Describe how a hemlock tree is important to life in the forest during each of the following stages in the life cycle.

A healthy hemlock tree:

____

____

An old hemlock tree with dead top and dying branches:

____

____

Old hemlock tree with soft wood and loose bark:

____

____

An old hemlock with old woodpecker homes:

____

____

Hemlock log on forest floor:

____

____

13. Explain how each of the following is important to a forest ecosystem.

Bats, woodpeckers and other birds:

____

____

Pileated woodpeckers (and other species of woodpeckers):

____

____

Nurse logs:

____

____

[page 22]

1. Place a checkmark beside the only animal that is <u>not</u> a predator.
   ___ earthworm ____ human ____ robin ____ fox

2. It is the job of the predator to hunt and kill animals for food. Being a predator is an animal's __________________.

3. Foxes are omnivores. As a predator they may eat ______________ or __________ __________________. As a herbivore they might eat __________________, ________________, or ____________________.

4. Explain why a fox may raid a farmer's chicken house. ______________________ ____

5. Most of the time, most predators are **(destructive/valuable)** to humans.

6. Assume that a pair of rats produces 2 offspring every three months. Three months after they are born, the young rats are sexually mature and breed. This means that every three months the rat population will double.

   A pair of rats moves into farmer Brown's barn. Just after arriving the female rat gives birth to two healthy babies. Farmer Brown stores the hog feed in open bins so the rats have access to all the food they can eat.

   Complete the following chart to show the total population of rats that will exist 24 months later if each pair of rats produces 2 healthy offspring and all rats survive.

Time	(Months)	3	6	9	12	15	18	21	24
# Rats	4								

   Two years after the pair of rats moved into the barn the rat population was ______.

7. **Think about it!** What limiting factors were missing in farmer Brown's barn?
   ____
   ____

8. Predators make the population stronger by ______________________________ ____.

9. Sometimes people are paid to kill predators. The money paid for each predator is referred to as a ____________________. Often money has been paid for those predators that are killing ________________ species.

10. In many cases when bounties were paid for predators, the population of the game species **(did/did not)** increase as expected.

11. Humans often make the coyote's job of killing easier by destroying the rabbit's __________________.

12. Predators often serve as an insurance policy against ____________________

13. **Think about it!** How might your life differ if there were no predators.

____

____

____

____

____

14. Make the connections between the vocabulary terms: bounty, game species, prey and predator. Add additional terms as needed to map the ideas presented in this case study.

Name ______________________________ Date ________ Class Period ____

Unit I — 2.4 The Case of the Missing Hawks

[page 24]

1. What caused the death of the pine trees surrounding the Reading reservoir? ______
____

2. What is the most important primary consumer in the meadow ecosystem?
____

   The population is normally controlled by ______________________, such as ____

3. The meadow mice normally eat seeds, berries, tender plant stems, and sometimes insects. Since tree bark is not a food in their normal diet. What series of events occurred that caused the mice to eat the bark of the pine trees at the Reading Reservoir? ____
____
____

4. The major food in the diet of the rough-legged hawk is the **(meadow mouse /rabbit /pheasant).**

   The diet was determined by examination of ______________________

5. Draw a food web using at least six of the organisms found in the area around the Reading reservoir.

6. The bounty offered for goshawks resulted in the killing of 76 goshawks that sometimes kill the farmer's ______________.

   Nearly 300 other species of hawks were mistakenly killed. These hawks would have helped the farmer by killing ____________________________.

   As a result, the bounty system **(helped/hurt)** the farmer.

Name ________________________________ Date __________ Class Period _____

Unit I — 2.5 Coyotes — an Opportunistic Omnivore

Introduction [25]

1. When the settlers first came, the major ecosystem in the Northeast was **(forest/prairie)** that was the home of this major predator **(coyotes/wolves)**.

2. Most predator populations have **(decreased/increased)** in size due to loss of ________________ and ____________________________________.

Return of the Coyote [25]

3. The population of coyotes in the eastern U.S. is **(decreasing/increasing)** probably due to the changing habitat caused by ______________________ of forests and elimination of these two predators, the ________________ and the ____________________________________.

4. Explain why an increase in the population of white-tailed deer may cause a decrease in the population of coyotes. ____
____
____
____

5. A good habitat for a coyote includes cover and a good supply of food. Describe what is needed by the coyote.:
Cover: ____
____
Food: ____
____

6. Draw a food web/concept map that shows the relationships between plants, the snowshoe hare, the white-tailed deer, and the coyote.

7. Explain how the large population of white-tailed deer is an indirect limiting factor for Pennsylvania's coyote population.
____
____
____
____

The Persistent Predator [26]

8. List several methods that ranchers use to control the coyote populations.
____

Name ______________________ Date ____________ Class Period _____

9. In the western part of the U.S. the population of coyotes is **(decreasing /increasing /stable).**

10. Killing coyotes on western ranches has caused the reproduction rate of the coyote to **(decrease/increase)**.

11. The major source of food for coyotes on western ranches is **(poultry /livestock /rodents)**.

12. The niche of a coyotes that feeds on a large animal like a deer or cow is probably a **(predator/scavenger)**. What evidence supports this choice.

____
____
____
____

13. Draw a food web shows the coyote is an omnivore. Write the niche (herbivore, prey, predator, or scavenger) indicating the role of the coyote on each arrow.

A New Approach [27]

9. Describe methods being used by ranchers today to reduce the loss of sheep to coyotes and list any disadvantages of these methods.

____
____
____
____

10. Give two ways that bonding sheep with cattle will increase a ranchers income.

____
____

Name ____ Date __________ Class Period ____

Unit I — 3.1 The Flow of Energy Through an Ecosystem

Introduction [29]

1. When sunlight strikes a grain of sand, a sidewalk or a building, it is either ____________________ or ____________________ by the objects.

   Most of the energy that is absorbed by a living or nonliving object is changed to ____________. If water is present, the most of the energy is used in the process of __________________________.

   The loss of water vapor from the openings in leaves is called ________________.

2. When sun shines on a green leaf, a small amount of the solar energy is absorbed and trapped by ____________________.

3. The trapped energy is used in the process called ____________________. In this process low energy molecules of ____________________ and ____________ are combined to form a high energy molecule of __________________.

4. Some of the sugar made during photosynthesis is changed into ____________-rich compounds including ____________, ______________ and ____________. An example of an oil is corn oil. Name some other oils produced by plants.

   ____

5. **Think about it!** Explain the statement: "Without chlorophyll there would be no food chains." ____

   ____

Building a Pyramid [29]

6. **Biomass** refers to the ______________ of all organic matter at any level in a food chain.

7. Which of the following will always have the greatest biomass? **(carnivores /herbivores /producers)**

8. At each level of a food chain, the biomass of organisms **(decreases/increases).**

9. Label the boxes below to represent the biomass of the following organisms: **carnivores (C), herbivores (H), producers (P).**

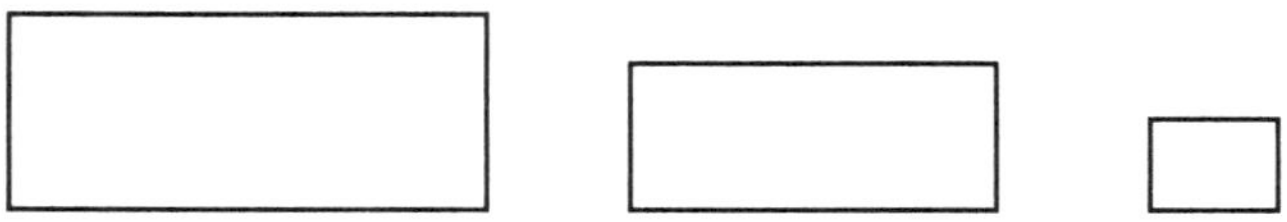

10. Draw and label a diagram that shows how biomass changes with each step in a food chain.

Name ________________________________ Date _________ Class Period ____

11. Most of the food made by a plant (producer) is **(stored in/used by)** the plant.

12. For each 1000 Calories of energy a plant absorbs from the sun, only ____ Calories will be available to a deer. Only ______ Calories of that energy becomes venison.

13. Most of the energy in the plants that are eaten by a deer will be **(changed to heat /used for daily activities)**.

14. The rabbits that Mr. Schultz raised provided ______________ for the greenhouse.

15. List three ways in which you use the Calories in the food you eat. _____________

____

16. The human digestive system cannot digest the walls of plant cells. Thus the energy in plant cell walls can not be used by humans. What happens to the energy in the cell walls of the raw carrots and lettuce served in the school cafeteria?

____

____

____

17. Only ______ percent of the energy in a plant eaten by a grasshopper becomes grasshopper biomass. List the ways that the energy in the plant becomes "lost" as it is passed through the food chain.

____

18. Because of this loss of energy the **(biomass/number)** of organisms in a food chain always forms a pyramid.

Numbers in a Pyramid [31]

19. The number of organisms in a food chain will not form a pyramid if the organism eaten is _____________ Will the number of organisms in the following food chain form a pyramid? ________

oak trees -------> gypsy moth larvae ---------> skunks

20. Which of the following diagrams best represents the number of organisms at each level in this food chain? _______

acacia trees ---------> elephants -----------> fleas

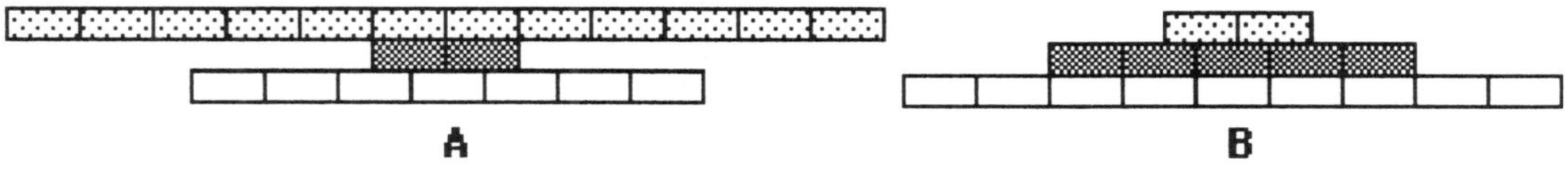

The Population Connection [31]

21. A plot of land can support more people if the people eat **(animals/plants)**.

22. Countries with little land and many people depend on food chains with **(few/many)** steps. This explains why traditional Chinese dishes contain **(few vegetables/small quantities of meat)**.

Name ____ **Date** __________ **Class Period** ____

Corn Fields and Coral Reefs [32]

23. **Productivity** refers to the _________________ of ______________ that can be produced per unit area (acre or hectare).

24. Rank the following ecosystems in order from greatest to least productive.
    coral reed _____
    corn field _____
    mountain meadow _____

25. Energy **(is/is not)** recycled in an ecosystem. Because of this the organisms in all ecosystems are dependent upon the ___________.

Try a little Math

26. Assume that 1% of the light energy a plant receives from the sun is changed into plant biomass. Assume that animals store 10 percent of the food energy as animal biomass. If a corn field receives 10,000 Calories of light energy:
    (a) How much energy is available to a person who eats corn? _____
    (b) How much energy is available to a person who eats beef? _____
    (c) How much energy is available to a person who eats frog legs? _____

Name ______________________________ Date ________ Class Period ____

Unit I — 4.1 Cycles in the Ecosystem

Introduction [34]

1. Energy **(is/is not)** recycled in an ecosystem. It is _______ by an organism or lost as __________. Energy from the ________ replaces the energy which is "lost."

2. The sun's energy is changed into a form that can be used by organisms in the process of ________________________.

3. Matter refers to the ___________________ that form or are needed by organisms in the ecosystem.

Nature Recycles [34]

4. Place a checkmark beside the statement that best describes matter and energy in an ecosystem.
   [ ] Both matter and energy are constantly recycled in the ecosystem.
   [ ] New supplies of both matter and energy are needed to maintain life in an ecosystem.
   [ ] Matter is recycled in an ecosystem, but energy must be constantly replaced.
   [ ] Energy is recycled in an ecosystem, but matter must be constantly replaced.

The Carbon-Oxygen Cycle [35]

5. Plants produce sugar and other compounds that contain carbon and are called ______________________. These compounds are a source of ______________ or serve as ___________________ for other compounds such as proteins and oils.

6. The organisms that are responsible for producing most of the oxygen in the atmosphere are **(algae/trees)**.

7. Oxygen is needed by both plants and animals for the process of ______________.

8. Organic chemicals with offensive odors are produced when respiration occurs without sufficient supplies of ________________.

9. Only two types of organisms can convert solar energy into chemical energy in the process called photosynthesis. Photosynthesis occurs only in _______________ and ______________.

10. The use of carbon dioxide by plants and algae is a **(fast/slow)** process. The release of carbon dioxide by **(combustion/respiration)** is a fast process. This is causing the level of carbon dioxide in the atmosphere to **(decrease/increase)**.

11. Indicate whether each of the following statements describes **photosynthesis (P)** or **respiration (R)**.

____ A. requires carbon dioxide
____ B. produces carbon dioxide.
____ C. requires light energy
____ D. manufactures sugar.
____ E. uses oxygen.
____ F. organic compounds are broken down.
____ G. occurs in both plants and animals.
____ H. decomposers break down dead organisms.
____ I. oxygen is a waste product.
____ J. process that is the most like combustion.

Name ______________________________ Date ________ Class Period ____

The Nitrogen Cycle [36]

12. Both animals and plants use nitrogen to make __________________..

13. The only organisms that can use nitrogen gas directly from the air are blue-green **(algae/plants)** and special types of **(bacteria/plants)**.

14. Organisms that change nitrogen gas into fertilizers that can be used by plants are called ______________________. Nitrogen can be used by plants only if it is "fixed" in the form of ____________ or ______________ ions.

15. Nitrogen-fixing bacteria live in the soil and in _____________ on the roots of plants called _____________. Plants that are members of the legume family include ____

16. Animals must get the nitrogen they need from ____________ or ____________ in the form of ________________.

17. The nitrogen cycle depends upon three kinds of bacteria.
Nitrogen-fixing bacteria make nitrogen gas into ______________ that can be used by plants.
Other bacteria (decomposers) change _____________ in dead plants and animals into fertilizers.
Still other special forms of bacteria change unused fertilizers into ____________ that is released into the atmosphere.

18. During thunderstorms, the intense heat of _______________ causes nitrogen in the atmosphere to combine with _______________.
Dissolved in the rain, the nitrogen (oxide) enters the soil where _______________ convert it into ______________ fertilizer.

19. Farmers add commercial ________________ or grow ______________ crops to increase nitrates in the soil before growing crops (like corn) that require large amounts of nitrogen.

20. In forest fires and motor vehicles, the high temperatures during the process of ________________ causes oxygen to combine with ______________ in the atmosphere. The ________________________ that are created may fertilize some crops.

The Mineral Cycle [37]

21. The source of minerals in most ecosystems is __________. ______________ and _______________ forces release minerals from rocks in a process called ____________________.

22. Explain how each of the following **weathering agents** remove minerals from rocks. Include both chemical and physical forces.
Wind: ____
Water: ____
Plant roots: ____
____

Name ______________________________ Date ________ Class Period ____

23.__________________ refers to the loss of minerals that are dissolved in water as it moves through the soil.
Humans actively remove rocks containing minerals in the process of __________.
Human actions also contribute to the removal mineral particles by wind and water in a process called ______________.

The Water Cycle [38]

25. Identify the process described.

______________ Most water is returned to the atmosphere by this process.
______________ Plants lose water vapor from openings in their leaves through this process.
______________ This process results in the formation of clouds.
______________ Moisture is returned to the earth by this process.
______________ This process increases the supply of ground water.

26. Identify the three ways in which water enters a stream.

____________________ ____________________ ____________________

Vocabulary Review

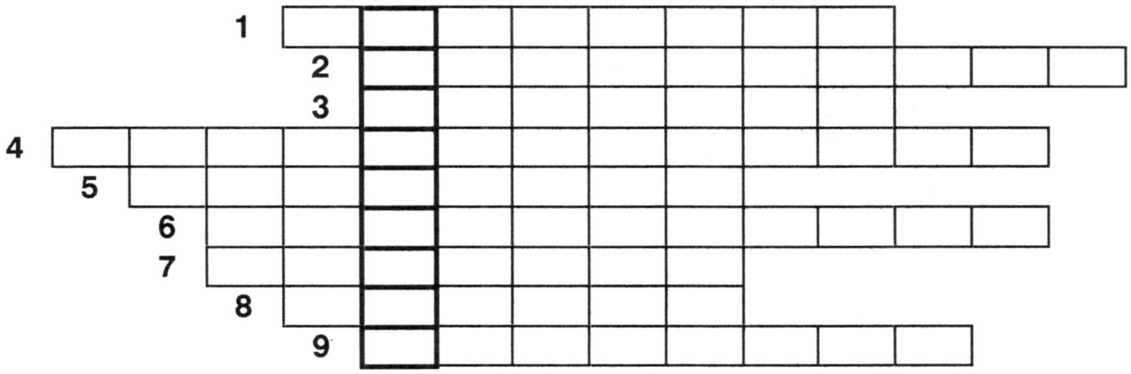

1 The process in which minerals are dissolved and carried away by water.
2 The process that involves the burning of compounds containing carbon and the release of carbon dioxide and nitrogen oxides into the atmosphere.
3 A compound that contains carbon.
4 The diffusion of water vapor from the openings in the leaves of plants.
5 Matter in an ecosystem is __________ but energy is not.
6 The process used by most living organisms to break down organic compounds.
7 A form of nitrogen that is used as a fertilizer.
8 A type of plant that has nitrogen-fixing bacteria on the roots.
9 Chemicals that are released from rocks by the action of weathering agents.

Name ______________________ Date ________ Class Period ____

Unit I — 5.1 Succession — Changing Ecosystems

[page 40]

1. **Succession** is a series of ___________ that occur in an ________________ with the passing of _____________.

2. The changes observed in succession are caused by the ____________________ that live in the ecosystem. One example of change in an ecosystem is the ________ created by the growth of trees.

3. <u>As you read</u>, (1) identify the stage of succession shown in each of the diagrams, and (2) identify the major types of plants that are present in each stage.

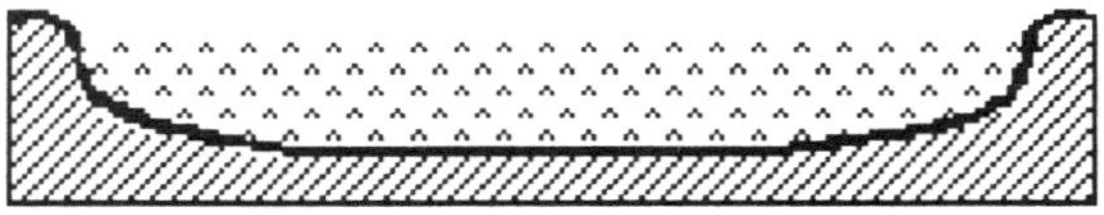

1 ____________________
2 ____________________
____

1 ____________________
2 ____________________
____

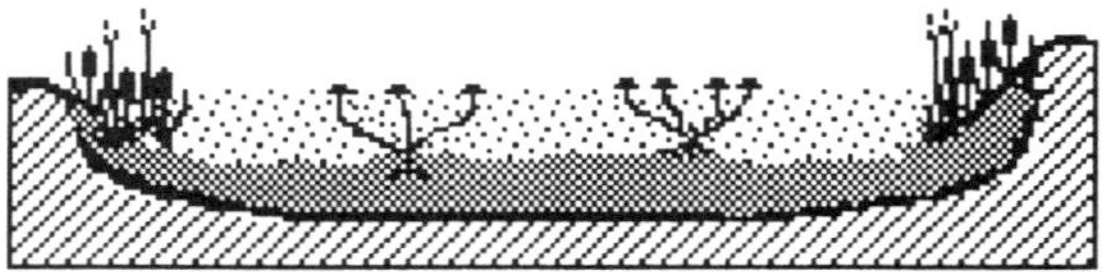

1 ____________________
2 ____________________
____

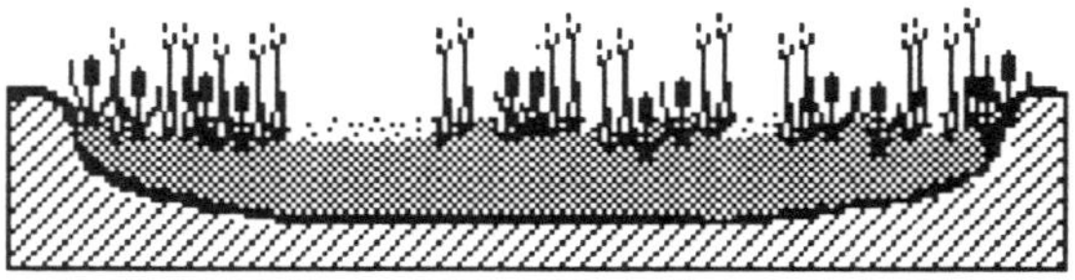

1 ____________________
2 ____________________
____

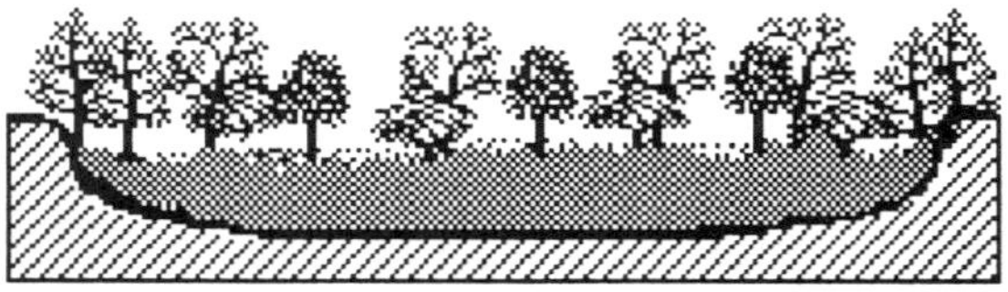

1 ____________________
2 ____________________
____

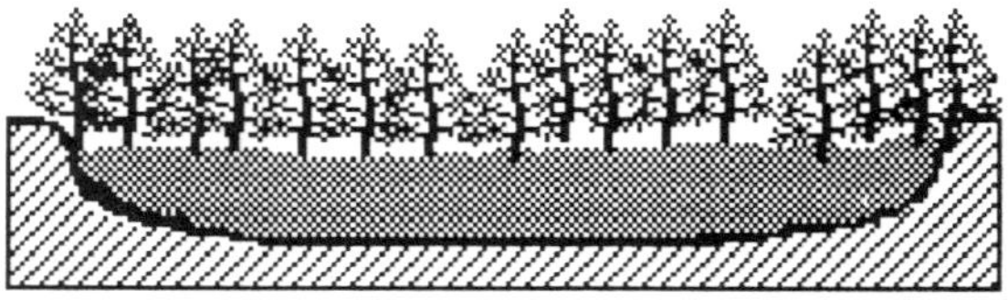

1 ____________________
2 ____________________
____

4. A turtle enters a pond and begins to move across the bottom. What will the turtle find if the pond is in the pioneer stage? __________________________________
____

5. When the bottom of the pond becomes covered with plants, species of fish that lived in the pioneer stage will disappear because the environment is no longer suitable and they stop __________________________ .

Name ______________________ **Date** _________ **Class Period** ____

6. Identify the <u>first</u> stage of succession in which each of the following organisms would be found.

___ A. caddisfly larvae with cases made from plants
___ B. caddisfly larvae with cases made of sand
___ C. dragonfly nymphs that climb to the surface
___ D. mussels that move across the bottom
___ E. fish that lay eggs on plant leaves
___ F. fish that nest on the muddy bottom
___ G. diving spiders
___ H. frogs and salamanders, but no fish
___ I. trees

(1) beech-maple forest
(2) emerging plant
(3) marsh
(4) submerged plant
(5) swamp
(6) pioneer

7. Arrange the stages listed in question 6 in the order in which they occur. Write the name of each stage on the line following the number.
(1) ______________________ (4) ______________________
(2) ______________________ (5) ______________________
(3) ______________________ (6) ______________________

8. Plants have invaded leaving no large open areas of water in the **(emerging plant stage/marsh)**.
A swamp is **(drier/wetter)** than a marsh.
Trees are present in the **(marsh/swamp)**.
A swamp is **(drier/wetter)** than a forest.

9. The level of dissolved oxygen **(decreases/increases)** as the pond ages.

10. On the Great Plains the stage of succession that often follows the marsh is the ______________. The growth of trees is prevented by: (1) ______________ or (2)______________________________

Name ______________________________ Date ________ Class Period ____

Unit I — 5.2 The Disappearing Lake

Introduction [43]

1. Large masses of ice that form from the accumulation of snow over many years are called ________________.

2. The weight of the ice and snow causes the glacier to move down slope and sometimes act as a __________________________ creating a ridge of rock.

Kettle Lakes [43]

3. The action of some glaciers creates a dam, and when the ice melts a __________ ______________ is formed.

4. Kettle lakes are unique because they have no ______________ or __________ that allow water to enter or leave the lake.

5. Which of the following is <u>never</u> a source of water for a kettle lake?
   ___ melting ice ___ precipitation ___ runoff (watershed) ___ streams

6. The best model of a kettle lake is a **(bathtub/cooking pot)**. The major way in which water is lost from the "kettle" lake is by the process of ______________ .

7. In the pioneer stage, the bottom of a kettle lake is covered with **(mud/rocks)**.

8. A kettle lake ages more slowly than other lakes because of the lack of ________ and ______________ carried into the lake by streams.

9. As the lake ages the level of oxygen **(decreases/increases)**.

10. As the lake ages the level of carbon dioxide **(decreases/increases)**.

11. The limiting factor for fish in the lake is the level of **(carbon dioxide/oxygen)**.

12. Identify the acid described in each of the following statements.

______________ A. formed when carbon dioxide combines with water.

______________ B. produced by your muscles during vigorous exercise (when there is an insufficient supply of oxygen).

______________ C. produced by bacteria breaking down plant material when the supply of oxygen is limited.

______________ D. gives the water an amber or "tea" color.

A Floating Mat [44]

13. Aging lakes with a flow of water become marshes and then, unless fire sweeps through the region or they are drained for farming, they become ____________ . Aging lakes without a flow of water become ____________.

14. The major limiting factors for the organisms in a bog ecosystem are the increasing level of ______________ and the decreasing supply of ______________.

Name ______________________________ Date _________ Class Period ____

15. Identify the plant described in each of the following statements.

__________________ A. trees that grow along the shore of lakes.

__________________ B. grass-like plants that emerge near the shore of the lake.

__________________ C. an evergreen plant that grows as a floating mat.

16. As the sphagnum moss grows, the lower part of the plant dies and sinks. It doesn't decay completely because the high level of __________ kills the ____________. The partly decomposed moss is called __________________.

17. Describe the major uses of peat moss in:
A. countries with abundant supplies, such as Denmark and Ireland: ___________
B. in the United States: ______________________________________

18. Identify the property of sphagnum moss that made it suitable for the following uses:
A. diapers for American Indian babies: ______________________________
B. bandages for American soldiers: ________________________________

The "Eye" of the Bog [45]

19. The area of open water in the center of the bog is called the _________ of the bog.

20. **Think about it!** Would you expect to find fish in the eye of the bog? _______
Explain. ____

21. Why is sphagnum moss considered to be the most important or dominant plant in the bog stage of succession? ______________________________________
____

22. Identify several plants that might be found on the sphagnum mat.
____

23. Two important limiting factors that prevent most plants from growing on the sphagnum mat are the high level of _____________ and low level of _________.

24. Plants that grow in the bog grow very slowly due to the lack of ______________. Some plants, such as the sundew and pitcher plant, absorb extra minerals by "eating" _____________.

The Invasion of Woody Plants [46]

25. Identify three small shrubs that grow on the thickened sphagnum mat.
____

26. Identify two larger shrubs found in bogs.
____

27. Identify two species of trees that are found growing in boreal bogs in the Northern U.S. (1) _________________________ (2) ____________________

Name ______________________ Date __________ Class Period ____

28. As the bog becomes drier these trees will be replaced by hardwood trees, such as ______________________ and the bog changes into a ____________.

29. Identify one species of animal that lives only in a bog. ______________________
Predict what will happen to the bog turtle as the bog changes into a forest.

____

____

____

30. In the diagram below, the center circle represents the "eye" of the bog. The largest circle represents the original shoreline of the kettle lake. Write the number of each plant listed below in the zone where it would most likely be found. Note: Some plants might be found in more than one zone.

1 black spruce trees
2 cranberry shrubs
3 sphagnum moss
4 lily pads
5 pticher plants
6 leatherleaf shrubs
7 bog cotton
8 poison sumac
9 tamarack trees
10 sundew plants
11 orchids
12 red maple

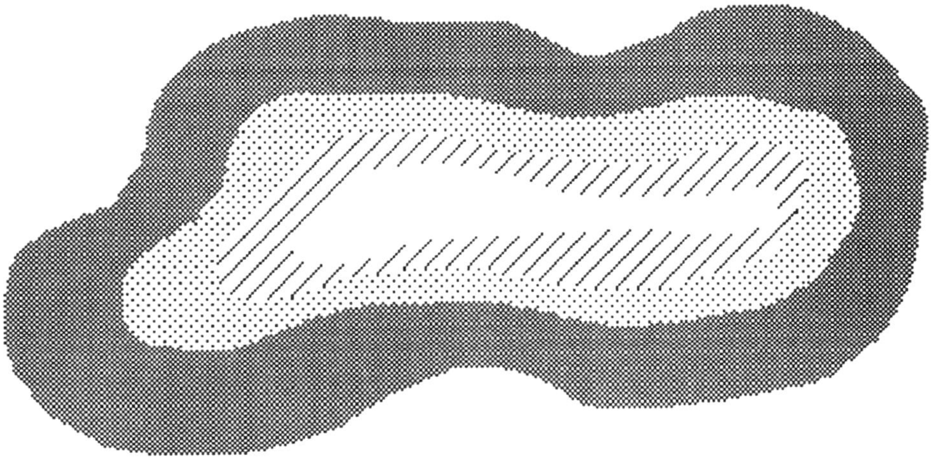

The Disappearing Lake

UNIT I — 6.1 More than Trees — an Old-Growth Forest Ecosystem

Introduction [48]

1. A forest is a complex ___________________.

2. The original forests ecosystems that have not been logged are also called ______________ forests or ____________________ forests because many of the trees are ancient.

3. Most of the original forests have been logged, only about ______ % remain.

Penn's Woods [48]

4. The conservation organization whose goal is to preserve the best examples of a wide variety of ecosystems is the **(Audubon Society/Nature Conservancy)**.

5. Are any virgin forests located in the state where you live? ___________
If so, where are they located? ____
Are they protected? ____________
What organization or government agency is responsible for them? ____________
____
[Hint: If necessary, call the state department in charge of natural resources to find this information.]

6. Write a sentence that reflects the feelings of each of the following groups of people about the forest:
American Indians: ____
____
Colonists: ____
____

7. When the colonists came to America one of the animals in the virgin forest ecosystem was the wolf. What children's stories have you read that reflect the colonist's fear of the wolf? ____________________________________
____

8. What was the major reason that trees in Pennsylvania's forest were cut during each of the periods listed below?
Before 1820: ____
After 1820: ____

9. The tree that was cut for its bark was **(hemlock/pine)**. The bark was used for **(making paints/tanning leather)**.

10. When land was sold, owners often retained **(bark rights/development rights)**.

11. In the mid-1800s the lumber-producing capital of the world was _____________.
More than 100 years later, in the late 1900s the lumber-producing capital is in the ____________________ region of North America.

Old-Growth Forests [49]

12. The dominant species of tree in the old-growth forests of the Pacific Northwest is the ____________________.

13. An old-growth forest is best described as a forest with **(a variety of/only one)** tree species of **(all ages/the same age)** and **(all sizes/the same size)**.

14. Give reasons why each of the following are important to a healthy old-growth ecosystem.
healthy mature trees: ____
____
dead and dying trees: ____
____
fallen trees: ____
____

15. Complete the following statements:
A dead tree in a forest is not wasted because it ______________________
____
A forest is more than a place with trees, it is a place ____________________
____

16. The greater the ____________________ of an ecosystem, the more ____________ the ecosystem.
What evidence is there that the old-growth forest has biodiversity?_____________
____

17. Mammals living in the old-growth forest that are at risk include the fisher, Sierra Nevada red fox, pine marten and wolverine. What is the niche of these mammals? __________________ Why do they need large areas of old-growth? ___________
____

18. Why does the seabird called the marbled murrelet need old-growth? ___________
____

19. What bird is an **indicator species** for an old-growth forest? ______________ ______________ What does the presence of an indicator species tell us about the ecosystem? ____

20. Match the organisms listed below with the correct statements. You may use more than one number for some statements.

(1) red-backed voles	**(4) other small mammals**
(2) Northern flying squirrel	**(5) tree roots**
(3) Northern spotted owl	**(6) truffles**

__________ A. controls population of small mammals in the old-growth forest.
__________ B. food for the Northern spotted owl.
__________ C. fungi that gets food from the roots of trees.
__________ D. source of food for red-backed voles.
__________ E. source of food for truffles.
__________ F. source of food for flying squirrels and other small mammals
__________ G. carries spores of truffles to areas where young trees are growing.

21. Draw a food web that includes the northern spotted owl. Remember that a food web must begin with a producer. Be sure that you can replace the -----> with the words "is (are) eaten by" or "provide food for."

22. Truffles have a symbiotic relationship with trees. A symbiotic relationship might be described as a "you help me and I'll help you" agreement.
Scientist say that truffles are essential to trees because the truffle helps trees:
____
____

Trees are essential to truffles because the trees: ____________________
____

23. Select the word that best describes how the population will change in each situation described. Be prepared to justify your answer. **D = decrease** **I = increase**

_______ If more old-growth forest is cut, the flying squirrel population will:
_______ If the old-growth forest is cut, the truffle population will:
_______ If the truffle population decreases, the small mammal population will:
_______ If the small mammal population decreases the owl population will:
_______ If the flying squirrel population decreases, the owl population will:

24. Predict how the above populations will change in a section of old-growth forest that is saved, but is not large enough to support a population of northern spotted owls.
____
____
____

25. What is the relationship between the old-growth forest and the populations of trout and salmon in the northwest rivers? ____________________
____
____

Multiple-use Management [51]

26. Of the original virgin forests is in **(National Forests/National Parks)** that are managed by **(conservation organizations/the federal government)**.

27. Place a **P** beside the area(s) where trees are protected and can not be logged.
___ National Forest ___ National Park ___ designated Wilderness Area

28. By law, the U.S. Forest Service can sell permits to log most of the remaining old-growth forest. **(true/false)**

29. National Forests must be managed for "multiple uses." Place a checkmark beside those activities which are allowed in a National Forest.

____ fish and wildlife habitat	____ recreation
____ grazing	____ timbering (logging)
____ mining	____ watershed protection

30. The uses permitted in a National Forest affect the local ____________________ .

31. Oil companies may be permitted to "mine" the oil and gas deposits under a National Forest. Lead, gold, and other mineral resources may be mined within the boundaries of a National Forest. What negative impacts might these activities have on the other uses of a National Forest? ____________________________________

____

32. According to critics, what use of National Forests is often given priority over other uses? ______________________________. Who makes the final decision about the amount of logging in National Forests? ____________________________

33. Refer to the glossary for the definition of the term **renewable resource.** According to this definition, are virgin forests a renewable resource? _______ Are old-growth forests a renewable resource? _________ If so, how must they be managed to be renewable? ____

____

Conflict in the Northwest [51]

34. Place a checkmark beside the position held by each of the following groups concerning the issue of cutting the old-growth forests and give one reason that supports their position.

Environmentalists/Ecologists: ____cut ____save
Reason: ____

____

____

Foresters: ____cut ____save
Reason: ____

____

____

Loggers and millworkers: ____cut ____save
Reason: ____

____

____

School board member: ____cut ____save
Reason: ____

____

____

Businesses in logging towns: ____cut ____save
Reason: ____

____

____

Fishermen: ____cut ____save
Reason: ____

____

____

35. "Locking up" the forest and denying permits for logging is only one factor that is reducing the number of loggers and millworkers employed in the Pacific Northwest. List five other factors that are costing the loss of jobs in the timber industry.

____

____

____

How Much is Enough? [53]

36. What are the two requirements of the National Forest Management Act passed in 1976? ____

____

37. Do you think that old-growth forest ecosystems sustainable? ______________
Defend your position.____

____

____

38. Why are scientists concerned about the information from satellite images of forests in the Pacific Northwest? ____

____

39. Describe a management plan that is needed to sustain the old-growth forest ecosystem. ____

____

____

40. What are the requirements that must be met to provide a sustained-yield of old-growth lumber and maintain old-growth ecosystems? ______________________

____

41. Should sustainability of old-growth be the goal of forest management? __________
Defend your position.

____

____

____

UNIT I — 6.2 Logging — Making Clearcut Decisions

Introduction [55]

1. Cutting every tree in a selected area of the forest is known as **(clearcut/selective)** logging. Another term for cutting a tree is ________________.

Clearcut in a Mixed Oak Forest [56]

2. Match the animals listed below with their preferred habitat.
   CB—clearcut areas with young "brushy" growth
   MT—area with mature trees

_____ squirrels	_____ adult turkeys	_____ ruffed grouse
_____ rabbits	_____ young turkeys	_____ deer

3. Good management of a forest may include the production of timber but it should also include production of quality ____________________________.

4. A forester may decide that trees in a specific section of a forest should not be cut in order to protect _______________ areas or to protect _________________________.

5. Selected trees might be cut in an area of a forest where the goal is to create facilities for camping and picnicking. The management goal for this section of forest is ______________________________.

6. Although trees and wildlife are the most obvious resources in a forest, another resource may lie hidden below the forest floor. In these sections, the primary goal of forest management may be the ____________________________.

Managing for Wildlife [56]

7. In sections of forest that are managed for wildlife, some trees must be allowed to mature, grow old and die. Some species feed and nest in the ________________ created by the tops of the mature trees.

8. ______________ with hollow sections provide homes for cavity nesters. More than _________ species of birds need dead or dying trees.

9. Some species of wildlife depend upon the fruit and nuts or _________ produced by mature trees in the forest. Other species are grazers or browsers and depend upon ______________ that grows in clearcut areas.

10. Indicate the <u>minimum</u> percent of the forest that should be in each of the following stages if good forest management techniques are practiced.
    ____ old-growth ____ mature trees producing mast ____ brushy growth

If a Little is Good, is More Better? [57]

11. Clearcutting is good for those species that depend upon the forest for ___________ and the clearings for __________. The increase in the populations of those species that depend upon these conditions found at the edge of the forest is called the __________________________.

12. **"Deep forest species"** live in the ______________ of the forest. Clearcutting causes populations of these species to ______________.

13. ____________________ refers to the carving of an ecosystem into isolated tracts.

List four activities that divide ecosystems into isolated islands.

1 __________________ 3 __________________
2 __________________ 4 __________________

14. Describe the changes that may occur in an ecosystem that is fragmented.

____

____

____

15. There is **(greater/less)** biodiversity in fragmented ecosystems. These isolated (fragmented) ecosystems are **(less/more)** stable.

The Forest Cycle — Managing for Timber [57]

16. Compare the biodiversity of a forest managed for timber with the biodiversity of a natural forest. ____

____

17. The practice of growing only one type of vegetables, grains, or trees is called ____________________.

To Log or Not to Log — That is the Question [58]

18. List four specific situations where a permit to log trees should be denied.

1. ____

2. ____

3. ____

4. ____

The Harvest: Which Trees to Cut? [58]

19. Select the type of logging that is preferred in each of the following situations.

A. clearcut B. seed tree C. shelterwood D. selection

_____ 1. remove high-quality saw logs leaving smaller trees for later harvest.
_____ 2. remove trees from an "overmature" stand.
_____ 3. remove trees to create conditions needed by sun-dependent species and eliminate the need for replanting.
_____ 4. salvage trees after the eruption of Mt. Saint Helens volcanic blast.
_____ 5. remove poor quality trees to improve conditions for remaining trees.
_____ 6. remove trees to prepare for planting oaks or pines.
_____ 7. remove trees from an even-aged stand.
_____ 8. remove trees leaving some trees as a seed source and as protection for shade-tolerant species.

20. Give one advantage and one disadvantage of growing trees by natural regeneration.
Advantage: ____
Disadvantage: ____

21. Give one advantage of planting over natural regeneration.
____

22. Before replanting the forest it is often necessary to remove the debris on the forest floor. The debris includes **duff** [__________ and ______________] and **slash** the [____________ and __________________].

23. State two reasons why it is necessary to remove the debris on the forest floor before replanting. (1) ____
(2) ____
____

24. Identify the method of preparation for planting that offers the most advantages in each of the following situations listed below.
A. mechanical **B. slash burning**

_______ cost is the deciding factor
_______ soil is thin and there is only a shallow layer of duff
_______ slopes are steep

25. Give one advantage and one disadvantage of the use of herbicides.
Advantage: ____
Disadvantage: ____

26. **Think about it!** Plantlings are produced by tissue culture—a technique that produces genetically identical plants. What are the advantages and disadvantages of using plantlings to replant a forest?
____
____
____
____

27. Trees with the greatest commercial value are _______ and ____________. To insure that the best trees grow properly, poorly formed and overcrowded trees are removed in a selective cutting process known as ________________.

28. The rate of growth is increased by thinning and by the application of
________________________.

29. Forests must be protected from ____________, ____________ and ________________. The need for pesticides is greater in a forest with one species of tree because this ______________________ provides the pests with a much larger supply of food.

Forests Forever [60]

30. List four differences between an **intensively managed forest** and an unmanaged forest.
    1. ______________________________
    2. ______________________________
    3. ______________________________
    4. ______________________________

31. Why must we intensively manage forests? ______________________________
____

32. **Think about it!** Is there a connection between the world population and the need to intensively manage forests? ________ Explain. ______________________________
____

33. What is the significance of the slogan "forests forever"? ______________________________
____
____
____
____

Vocabulary Review

DOWN:

1 the cutting or felling of every tree on a specific tract of forest.
2 a dead tree that is still standing.
3 limbs, tree tops and other waste that remains after logging.
4 refers to the cutting of trees.
5 growing of only one plant species in a large area.
6 acorns and other nuts and fruits that lie on the forest floor.
7 layer of twigs and needles on the forest floor.
8 the cutting of trees so that the remaining forest ecosystem is in small isolated tracts.
9 the cutting of only certain trees such those that produce high quality saw logs.

ACROSS:

1 the area of the forest that is created by the tops of the mature trees.
2 young plants that are produced by tissue culture from selected parent plants.
3 plant material that is eaten by browsing and grazing animals.
4 chemicals applied to kill "weed" species.

Name ______________________________ Date __________ Class Period ____

Unit I — 6.3 Introduced Species

Introduction [63]

1. The **(grackle/starling)** was introduced to New York in 1890. Now this introduced species competes with native birds for ____________________ and ________________.

2. At times farmers may find starlings useful because they eat ____________, but at other times the farmers consider them a pest because they eat ______________.

3. Today there are rules for importing birds. One goal is to protect ______________ species and another goal is to prevent importing ____________ that could threaten the poultry industry and native species.

4. The two major reasons for the decline in some populations of parrots are: (1) ____________________ and (2) ______________________________.

Alien Plants [63]

5. Settlers brought some plants to America intentionally for use as ______________, __________________, or ____________. Others were accidentally imported in __________, ______________ or soil.

6. ________________ plants often can not compete with introduced or "naturalized" plants and the introduced plants replace the natives.

7. **Purple loosestrife, probably** introduced **(accidentally/intentionally)** in __________, has invaded ______________ and is **(decreasing/increasing)** the carrying capacity of these habitats by displacing native plants that provide ____________ and ______________ for wildlife.

8. Native plants in aquatic ecosystems along the Gulf Coast are being displaced by the ____________________________ that was imported from South America as an ____________________________.

9 Identify three methods that are being used to control the water hyacinth.
(1) ____
(2) ______________________________ (3) __________________________

The Little but Mighty Mollusk [64]

10. The tiny mollusk that recently invaded the Great Lakes is the ______________. It probably was imported from Europe in ________________ carried in the ship's ballast tanks.

11. List two ways that the exploding mollusk population can endanger fish populations.
(1) ____
(2) ____

12. List two problems that zebra mussels cause for humans.
(1) ____
(2) ____

Name ______________________________ Date ________ Class Period ____

Troubles for Trees [65]

13. Two native tree populations that have been devastated by introduced species are the American ____________________ and the American ___________.

14. The American ______________, once the dominant tree in the eastern forests, was devastated by a fungus called _____________________ that was probably imported on ___________________________.

15. The fungus doesn't kill the tree's ____________, and scientist hope to save the American chestnut from extinction by developing a _______________ control from a __________________ or "hypovirulent" form of the fungus.

16. The fungus that kills the American elm is called ______________________. Probably imported on __________ from Europe, it is spread from tree to tree by the ____________________________.

17. Efforts to save the American elm include injecting the tree and soil with _________________ and _______________________.

18. Like the American chestnut and elm trees, the American Indian did not have any __________________ that would have protected them from introduced diseases.

19. Like the American Indians, Brazil's Yanomami people population declined in the 1980's. The increased death rate was due to diseases such as _______________. The diseases were introduced to the Yanomami people by _______________.

The Hitchhiker [66]

20. The most serious pest in the forests of the Northeast today is the ____________. The purpose of importing this species was for breeding an insect with an improved ability to produce __________.

21. The gypsy moth larvae is known as the hitchhiker because it is carried long distances by _________________. The gypsy moth found in Sequoia National Park was probably carried there by **(campers/the wind)**.

22. The gypsy moth prefers to breed on __________ trees. The larvae **(feeds only on oak leaves/feeds on many different trees)**.

23 From 1981 to 1983 the number of acres of forest affected by the gypsy moth **(decreased/increased)**.

24. The limiting factor that caused the decline in the gypsy moth population was ______________.

25. There is often a population explosion when an insect is introduced to a country because this limiting factor is often missing **(food supply/predators and parasites)**.

26. Give two causes of the decline in the gypsy moth population. (1)____________ ______________________, and (2)______________________________.

Name ______________________________ Date ________ Class Period ____

27. Describe the appearance of a forest three years after an invasion by a large population of gypsy moths. ______________________________
____

28. Indicate whether the changes described in question 27 will cause each of the following populations to **decrease (D)** or **increase (I)**.
____ white-footed mice ____ rabbits ____ squirrels
____ ruffed grouse ____ gypsy moth ____ deer

The Flight of the Bees [68]

29. A scientist in Brazil imported wild bees from **(Africa/Europe)**.

30. The scientist wanted to breed the African and European bees to produce a bee that was better adapted to **(pollinate crops/produce honey/a tropical climate)**.

31. The bees that are moving through Central America and into the United States are **(African/Africanized/European)** honey bees.

32. Honey bees **(are native/were introduced)** to the United States.

33. Place a + beside each statement that is true and a – beside each false statement.

___ Africanized bees are easily identified.
___ The Africanized bees have been labeled killers by the movie industry, but a sting from a European honey bee is as dangerous as a sting from an Africanized bee.
___ The Africanized bees are aggressive and unpredictable, and they have caused the death of humans and other animals.

34. List two ways in which the Africanized bees may affect the agricultural industry.
(1) ____________________ (2) ____________________

35. Beekeepers are using chemicals to control **(Africanized bees/parasitic mites)**.

36. Scientist imported bees from ______________. The advantage that these bees offer to beekeepers is that they are ______________.

The Medfly Attack [69]

37. The "Medfly" refers to this serious pest, the ______________. The adult fly deposits its eggs under the skin of 260 kinds of ______________.

38. Each time the fly is discovered, ______________ are imposed to prevent its spread and the government spends millions of dollars to protect __________ that are worth billions of dollars.

USDA Animal and Plant Health Inspection Service [69]

39. List two ways that imported pests cost Americans money. (1) ______________
______________ (2) ______________

40. What is the primary goal of the USDA Animal and Plant Health Inspection Service?
____

Name ________________________________ **Date** __________ **Class Period** ____

41 If you are returning to the United States from another country, you face a heavy fine if you do not declare these items. ______________________________

____

42. You set out on an around the world trip and you plan to bring home a handful of soil from each country you visit. When you return home will you be allowed to keep the soil? _______ Why? ______________________________.

43. You visit an uncle in Italy and enjoy olives that he grows. You bring home a small olive tree. Will you be allowed to keep your olive tree? ______________________

____

44. Importing a plant carries a risk of introducing a pest. Quarantines are not totally effective, why does APHIS allow any plants to be brought into the country?

____

____

____

Know Before You Go [70]

45. Some people have received heavy fines for trying to import a plant or animal. How can you avoid fines and possible loss of the plant or animal you are importing?

____

46. Dogs entering England must be quarantined for six months in order to avoid the introduction of the disease ______________.

Name ______________________________ Date ________ Class Period ____

Unit I — 6.4 Will Endangered Species Become Extinct?

Introduction [72]

1. If a species chances of survival and reproduction are in immediate jeopardy, the species is considered to be ________________. If a species is likely to become endangered it is classified as ________________. Once the species no longer exists on Earth it is ____________.

2. The government agency that is responsible for identifying and protecting species that are endangered is the United States Department of ______________.

3. According to the Endangered Species Act species are valuable for the following reasons. Give an example of a species that you think fits each description.
   esthetic ____
   ecological ____
   educational ____
   historical ____
   recreational ____
   scientific value ____

4. The Endangered Species Act protects species by making it illegal to __________, __________, or ________ species that are listed as endangered or threatened. It also makes it illegal to take their ______________.

5. Place a checkmark beside the statement which best describes the major goal of the Endangered Species Act.

___ to maintain examples of endangered species in zoos and nature centers for purposes of education.
___ to restore populations of species so they are self-sustaining member of their ecosystem.
___ to preserve genetic messages (DNA) for scientific research.

6. The Endangered species act requires that a ______________ plan be developed for each species listed as endangered or threatened.

7. The number of endangered species is **(decreasing/increasing)**.

8. The factors that limit the number of species added to the endangered species list each year are the ____.

Extinction: With or Without Humans [72]

9. Describe two possible events that may have led to the extinction of dinosaurs.
   ____
   ____

10. What group of animals survived the events that caused the extinction of the dinosaurs but is in jeopardy today? ________________
    According to the National Academy of Sciences, what is the major cause of the decline in populations of these animals? ____________________

Name ______________________________ Date ________ Class Period ____

11. Place a checkmark beside the statement(s) that is/are true.
___ Extinction is solely due to human activities.
___ Extinction is a natural process.
___ The rate of extinction is increasing due to human activities.

12. Place a checkmark beside each country in which human population growth is causing populations of some species to decline.
___ Africa ___ Asia ___ India ___ North America ___ South America

13. List three reasons why the big cat populations (lion, tiger, leopard, and cheetah) are declining. (1)____________________ (2) ________________________
(3) ________________________

A Little Fish vs. a Big Dam [73]

14. The habitat of the snail darter is a **(fast-flowing stream/lake)**. Building a dam would **(destroy/improve)** the snail darter's habitat.

15. The Endangered Species Preservation Act prevented construction of federal projects that would ____.

16. Place a checkmark beside the statements that correctly describe the situation involving the Tellico Dam.
___ The Tellico Dam project was canceled due to the presence of the snail darter in the Little Tennessee River.
___ A senate committee approved the dam because the need for low-cost electricity was more important than the value of the small fish.
___ The Congressional panel found that the Tellico Dam was not needed.
___ The Tellico Dam project was included in another senate bill and the dam was declared exempt from the Endangered Species regulations.

17. Place a checkmark beside the statement that correctly describes the snail darter's status.
___ The snail darter was successfully transplanted to other streams.
___ The snail darter was discovered in other streams.
___ The snail darter is extinct.
___ The snail darter is no longer listed as endangered.

Will the Warblers be Next? [74]

18. When populations of passenger pigeons were severely reduced in number, the remaining animals did not **(eat/mate)**.

19. The major causes of the decline of the passenger pigeon population were loss of habitat and unregulated hunting. List the major reasons for the decline of the Kirtland's warbler population.
(1) ________________________ (2) ________________________

19. The species that is extinct is the **(Kirtland's warbler/passenger pigeon).**

20. The Kirtland's warbler nests only under **(mature/young)** ______________ trees. As the trees grow, ______________ becomes a limiting factor.

Name ______________________ Date ________ Class Period ____

21. To be acceptable nesting habitat for the Kirtland's warbler, the soil type must be ________________ and the stand of young trees must be at least _______ acres.

22. Better control of forest fires **(decreased/increased)** the warblers habitat.

23. The bird that lays its eggs in the nest of other birds, including the Kirtland's warbler, is the ____________. Explain how this bird limits the number of young warblers that will survive? ____
____

24. As part of the recovery plan for the Kirtland's warbler, the U.S. Fish and Wildlife Service is ______________________________ the cowbirds.

The California Condor [75]

25. What is the niche of the California condor? ____________________

26. List three reasons why the California condor was shot by the early settlers.
(1) ____
(2) ____
(3) ____

27. How did the breeding pattern of the condors affect the price of condor eggs?
____

28. List three possible factors that might explain the "lack of breeding" noted in the 1972 condor study. (1) _________________,(2) _________________,
(3) ____________________

29. The condors were poisoned when they ate. (1) ___________________ in animals that had been shot and (2) __________________________________

30. The major reason for the decline in the number of young condors was **(broken egg shells/lack of mating)**.

31. As a part of the recovery plan for California condors, all condors were captured and became a part of a _____________ program that increased the number of condors. In 1991 scientists ___________________ condors at Los Padres National Forest.

Back from the Brink [76]

32. Give three reasons why massive numbers of bison were killed.
(1) ____
(2) ____
(3) ____

33. What steps were taken to save the bison from extinction?
(1) ____
(2) ____

34. Today bison are **(common/endangered/extinct)**.

Name ________________________ Date ________ Class Period ____

35. Ranchers are concerned that buffalo will spread the disease **(brucellosis/rabies)** to their cattle.
Although the disease doesn't affect buffalo, it causes cattle to ____________ their calves. How are the cattle being protected from this threat? ____________________
____

Return of the Wolves [77]

36. List three reasons why the wolf population declined. (1) __________________
(2) ______________________(3)________________________

37. What state supports a healthy population of wolves that can be legally hunted? ______________ The gray wolf is listen as threatened in ______________.
In all other states, the wolf is listed as __________________.

38. A captive breeding program has increased the number of ________ wolves.
Breeding pairs of this wolf species have been successfully reintroduced in ______________________ National Park.

39. Would you vote for the recovery plan that includes reintroducing 10 pairs of gray wolves into Yellowstone National Park? ___________ Support your position or if you are unsure, list the information you need before making a decision.
____
____
____

Why Not Let Them Die? [79]

40. Do you think that any of the species discussed in this section will become extinct? ____________ If so, which one(s)? ___________ What will be the cause of the extinction? ______________________________________

41. Why is it important that we save plants like the:
A. rosy periwinkle ____________________________________
B. wild sugar beets ____________________________________
____

42. Explain why we need a "library of living things." ____________________
____

43. Explain how removal of a species from an ecosystem might affect the ecosystem.
____
____

44. Species that keep an ecosystem in balance are called ______________ species.

45. The most critical threat to most species of wildlife today is the ______________
____

The Future of the Endangered Species Act [79]

46. If the Human Protection Act was passed, decisions regarding an endangered species would consider the _____________ value of a species.

Name ______________________________ Date ________ Class Period ____

47. An amendment to the Endangered Species Act allows "incidental taking" of individuals if it is a part of a ______________ plan.

Describe the action taken to help save these endangered species.
A. Desert tortoise: ____
B. Ridley sea turtle: ____

48. To save species we must save ________________.

Name ______________________________ Date __________ Class Period ____

Unit I — 6.5 A Question of Hunting

Introduction [82]

1. In the early 1930's many species of wildlife that are native to America were **(disappearing/abundant)**.

Money for Wildlife [82]

2. The 1937 Pittman-Robertson Act imposed a tax on ____________________ and ____________________. An amendment added during the 1970s placed a tax on ____________________ and ____________________.

3. Taxes established by the P-R Act must be used for ____________________.

4. What agency or department in your state is in charge of decisions concerning wildlife? ____

5. Most of the money used by the state wildlife conservation agencies comes from **(fees and taxes paid by hunters/general taxes paid by all people)**.

6. Fish and game agencies or conservation agencies that receive funds from the P-R Act must spend the money for wildlife. List three ways in which the money might be used.
(1) ____
(2) ____
(3) ____

7. Land that is a part of the National Wildlife Refuge System was purchased with money paid by **(hunters/taxpayers)**. The Refuge System provides habitat for **(many species/only migratory birds)**.

8. Most people who visit land that is a part of the National Wildlife Refuge System **(are/are not)** hunters.

Game vs. Nongame [83]

9. **Game species** are those species that **(are protected/can be legally hunted)**.

10. Wildlife management plans often favor **(game species/nongame species.** Some people feel that more species would benefit if wildlife were managed by **(conservation agencies/game agencies)** and if those agencies were supported by **(general taxes/hunting licenses)**.

11. **(Few/Most)** states support the activities of conservation departments with taxes collected as a general sales tax.

12. In your opinion, how should wildlife management plans be funded? ____________
____
____

Name ______________________ **Date** ________ **Class Period** ____

The Return of the Whitetails [83]

13. The white-tailed deer population in several states declined in the early 1900s due to ______________________ and loss of ____________.

14. The population of white-tailed deer in North America is more than ____________. The recovery is due to ____________ hunting; ______________ of game laws; and restoration of ______________.

15. Identify the problem that exists at Gettysburg National Military Park.

____

____

____

What recommendation was made at the end of the three-year study?

____

____

16. List one possible benefit of hunting for each of the following:
Hunter: ______________________________________
Deer population: ______________________________

Nature's Way [84]

17. The population that a given area of habitat can support is the ____________.

18. The largest deer will be found in areas **(with/without)** farming activity and areas that **(are/are not)** open to public hunting.

19. The number of deer producing twins will be greater in areas **(with/without)** farming activity and areas that **(are/are not)** open to public hunting.

Game Management [84]

20. What species is the major predator of white-tailed deer? ______________

21. What is the primary method of wildlife management? ______________

22. In order to determine the number of white-tailed deer that should be "harvested" the wildlife biologist or game manager must know:
(1) ______________ (2) ______________

23. Most wildlife that die of "natural causes" will die during the ________ season. List possible causes of death: ______________________________

24. The Michigan study showed that a population of grouse that was protected was **(larger than/the same size as/smaller than)** a population of grouse in an area with regulated hunting.

25. Hunting seasons have sometimes been canceled because of war or danger of fire during periods of drought. If the hunting season for white-tailed deer is canceled, the number of "road kills" will be expected to **(decrease/increase/remain the same)**.

Name ________________________________ Date ________ Class Period ____

Doves in Indiana [85]

26. **(Few/Most)** states have a hunting season for mourning doves.

27. Choose the correct word or phrase that reflects the information provided by the Indiana Fish and Wildlife Division during the legislative debates.

 (Most/Few) doves nest during the hunting season.
 (Both parents/Only the female dove) cares for the young.
 The mourning dove feeds **(mainly on ragweed seeds/on many types of seed)**.
 Mourning doves are mainly **(bones and feathers/breast muscle)**.
 In states with hunting seasons, populations of mourning doves have **(decreased/increased)**.

28. List the differences between the mourning dove and the passenger pigeon.

	Mourning Dove	Passenger Pigeon
Food		
Nesting Habitat		
Offspring		

To Hunt or Not to Hunt [86]

29. ________________ refers to any form of illegal hunting. The two major reasons why poachers kill wildlife are ________________ and ________________.

30. **(Poaching/Regulated hunting)** threatens certain wildlife populations.

31. The biggest threat to waterfowl populations today is **(destruction of the environment/regulated hunting)**.

32. If regulated hunting is not permitted, more animals will starve during the winter season. Give at least one advantage of each situation:
 An animal is killed by a hunter's gun: ________________________________
 ____

 An animal dies of starvation: ________________________________
 ____

33. In addition to hunting, wildlife management must consider ________________
 ____

34. You are the manager of the 185 acre Schlitz Audubon Center in Milwaukee. Here is the problem facing you.

 The carrying capacity of the nature center is about twelve deer. During the last ten years, the population of deer at the nature center has increased from fourteen to fifty. Twenty-seven fawns were born this spring.

Name ______________________________ Date ________ Class Period ____

The deer are damaging the plants at the nature center. A survey of the sixty neighbors who live closest to the center revealed that ninety-eight percent had deer visiting their yards. Seventy-seven percent of the neighbors had damage to their plants.

A few of the neighbors like having the deer in their back yards. One person built a feeder for the deer, but most of the neighbors are demanding that something be done. As manager of the nature center, you must present the possible solutions to the board of directors for a vote.

Give possible solutions, their advantages and disadvantages.

Solution #1: ______________________________

Advantages: ______________________________

____

Disadvantages: ______________________________

____

Solution #2: ______________________________

Advantages: ______________________________

____

Disadvantages: ______________________________

____

Solution #3: ______________________________

Advantages: ______________________________

____

Disadvantages: ______________________________

____

Solution #4: ______________________________

Advantages: ______________________________

____

Disadvantages: ______________________________

____

Name ______________________________ Date ________ Class Period ____

Unit II — 1.1 The Thin Blue Blanket

Introduction [91]

1. The mixture of gases that creates a thin blue blanket surrounding planet Earth is the ________________. The blanket is 78 percent __________ and 21 percent ______________.

2. The force that holds the blanket of gases close to the Earth is ______________.

3. Explain why Mercury has no atmosphere. ______________________________
____

4. The gas that helps create the hot climate on Venus is ____________________.

5. Identify two environmental conditions that make life as we know it impossible on Uranus. (1) ______________________________
(2) ______________________________________.

6. Using an apple as a model for the Earth, the atmosphere is represented by the apple's __________.

7. At higher altitudes the Earth's atmosphere is ______________. Explain how this characteristic affects mountain climbers. ______________________________
____.

The Ozone Shield — a Natural Sunscreen [92]

8. The two sources of energy that split oxygen molecules in the atmosphere are ____________________ and __________________.

9. The molecule formed when three oxygen atoms join is ______________. The symbol for this molecule is ________.

10. The layer of the atmosphere nearest the Earth is the ____________________. Within the upper layer, the ___________________, contains a protective shield of __________.

11. The concentration of ozone in the ozone shield is measured in parts per ________________ by volume. Ozone is constantly being created by the action of the sun's __________________ and destroyed by the action of ____________ __________ that is produced by volcanoes.

Biological Effects [92]

12. The ozone shield protects life on planet Earth from harmful ______________.

13. UV radiation causes these conditions in humans. (1) __________________
(2) ____________________. It may also interfere with the ability of the body's __________________ system to fight diseases.

14. Ultraviolet light **(is/is not)** necessary for photosynthesis. Increasing UV radiation **(decreases/increases)** the rate of photosynthesis in some plants.

Name ______________________________ Date ________ Class Period ____

When UV radiation is increased, crops that we use for food grow **(faster/slower)** and the yield is **(lower/higher)**.

15. There is evidence that increasing UV light harms ____________________ that is the basis of the ocean's food chain. There is also evidence that UV light interferes with the development of ___________ eggs.

CFCs—The Risks and Benefits of Technology [94]

16. Compounds that contain carbon, fluorine and chlorine are called CFCs or ______________________.

17. Identify three characteristics that made CFCs ideal for use in consumer products.
(1) ______________________ (2) ______________________
(3) ______________________.

18. **What do you think?** From your point of view, classify the following uses of CFCs as essential (E) or nonessential (N) uses.
____ blowing agents to make foam for insulation and packaging materials.
____ solvents for cleaning electronic parts.
____ air conditioning in your car
____ cooling system in refrigerators
____ propellants in aerosol sprays

19. What exception was made when CFCs were banned as propellants in aerosol sprays? ____

20. CFCs are **(<u>stable</u>/unstable)** in the troposphere. UV rays in the stratosphere split ________________ from CFCs. One chlorine atom may destroy 100,000 ____________ molecules. Eventually the chlorine atom returns to the lower atmosphere and washed out by ____________.

Warning Not Heeded [94]

21. When did scientists first discover evidence that linked CFCs to the destruction of the ozone layer? _______ When was the first action taken to stop the destruction of the ozone layer? _______ What use of CFCs was banned by the U.S., Canada and some European countries? ______________________________

22. Why weren't other uses of CFCs banned in 1977? ______________________
____

23. What chemical provided evidence that CFCs were causing the ozone "hole"?
____

24. Describe what is happening to the ozone shield. ______________________
____

25. Where is the loss of ozone the greatest? ________________ During what season does the greatest loss occur? __________ What factor contributes to the loss? ____

26. Is there really a hole in the ozone shield? _______ Explain. ______________
____

Name ______________________________ Date ________ Class Period ____

27. Since we live in the northern hemisphere, should we be concerned about the loss of ozone? ______________________________

The Montreal Protocol [95]

28. What is the major cause of ozone destruction? ______________________
What two steps must be taken to limit the destruction of the ozone shield?
(1) ______________________________
(2) ______________________________

29. What was the original goal of the Montreal Protocol on Substances that Deplete the Ozone layer? ______________________________
How did the 1990 amendment change the goal? ______________________
____

30. Why was an international agreement necessary? ______________________
____

New Technology [95]

31. What chemicals are replacing CFCs in refrigerators and air conditioners? ________
What makes these chemicals safer? ______________________
____

32. What characteristics are desirable for a chemical that replaces a CFC compound?
(1) ______________ (2) ______________
(3) ______________ (4) ______________

33. Are CFCs essential for cleaning electronic parts? ______ If not, why do you think that they were used for this purpose? ______________________
____

34. What is the essential use of halons? ______________________
In some situations, halons are being replaced by carbon dioxide. Why can't halons be replaced by carbon dioxide in all situations? ______________________
____

35. Describe one way that you can help protect the Earth's ozone shield
____
____

36. Describe how you would protect a small child from UV rays.
____
____
____
____

Name ______________________________ Date _________ Class Period ____

Unit II — 2.1 Weather Patterns

Weather Predictions [98]

1. ____________________ is the study of the physics and chemistry of the atmosphere. ________________ refers to atmospheric conditions that result from the interactions between components of the atmosphere.

2. Identify the four components of the atmosphere that affect our weather.
(1) ____________________, (2) ____________________,
(3) ____________________, (4) ____________________.

3. List three careers with an income that is dependent upon the weather.
(1) ____________________, (2) ____________________,
(3) ____________________, (4) ____________________.

4. Signs of nature that occur because of ______________ changes in the atmosphere accurately predict **(long-range/short-range)** weather.

5. During fall or spring nights, the chance of frost is increased if the sky is **(clear /cloudy)** because the Earth loses more heat without a blanket of ____________.

6. Before a storm the barometric pressure is **(low/high)**. If you see smoke rising straight from a chimney, you can predict a **(clear day/stormy weather)**.

7. An increase in static electricity before a storm may cause your cats to **(groom their coats more frequently/sleep more than usual)**.

8. While on a hunting trip in the high country of the Rocky Mountains, your guide notices that the elk have started to migrate earlier than usual. Should you break camp? ______ The elk are probably responding to a change in ____________ ____________________________. What may happen if you decide to stay? ____

9. The brown band on the woolly bear caterpillar is not a true indicator of coming winter weather because it's width is determined by **(environmental/genetic)** conditions.

The Weather Forecast [99]

10. Describe the role of each of the following technologies in weather forecasting.
Satellites: ____
Supercomputers: ____________________________________

11. Identify the types of information that satellites provide about atmospheric conditions. (1) ______________ (2) ______________ (3) ____________

12. What two advantages do satellites offer to weather forecasters. (1) ____________ ________________ (2) ________________________________

13. What advantage do weather balloons offer over satellites? ______________ ____

Name ______________________________ Date __________ Class Period ____

Storm Warnings 100]

14. The weather forecasters use Doppler radar to detect the formation of ____________ within a storm cloud.

15. What historic event led to the formation of the National Hurricane Center?
____

16. Today prediction of a hurricane's path has improved due to better information from ______________________, ____________, and specially equipped aircraft. This information permits advanced warning that reduces the **(loss of life/property damage)**.

Improving the Forecast [100]

17. Meteorologists still have much to learn about the movement of a river of air called the ____________ and the warming of water in the East Pacific called ____________.

18. It is difficult to make accurate long-range weather forecasts because of the number of ______________ that control weather.

19. Improving long-range forecasts may depend upon a better understanding of the link between weather patterns and the temperatures of __________________.

The Hydrogen Furnace [100]

20. Is the sun likely to run out of fuel in your lifetime? _______

21. Solar energy is the result of a process called _____________ in which the fuel is _______________ and the low-energy "waste" produced is _____________.

22. The driving force behind all weather is ____________ energy that is converted by the atmosphere to ____________.

The Air Around Us [101]

23. The transfer of _________ or ___________ waves of energy is called **radiation.**

24. As the waves travel through the atmosphere, light waves are scattered by __________________ and ________________. Beautiful sunsets and rainbows are due to the scattering of certain wavelengths of light. Would we see sunsets or rainbows if the earth had no atmosphere? ______

25. Give the correct percent of solar radiation that is:
A. reflected by particles, water vapor, or the earth. ____ percent.
B. absorbed by the atmosphere. ____ percent.
C. absorbed by the earth and objects on it. ____ percent.

26. When solar energy strikes an object, the energy that is absorbed is changed to __________ energy. The atmosphere is warmed by energy from the **(earth's surface/sun)**.

Name ______________________________ Date ________ Class Period ____

27. Daytime temperatures in a desert are extremely high because _____ percent of the solar energy reaches the Earth and is changed to heat. Nighttime temperatures get cold because deserts have very little ______________ to trap the heat.

28. In humid climates water vapor acts as a **(blanket/sheer curtain)** that reflects light energy during the daytime, but at night water vapor acts as a **(blanket/sheer curtain)** that prevents heat from escaping.

    Daytime temperatures in humid climates will be **(cooler/warmer)** than daytime temperatures in desert climates.

    Nighttime temperature in humid climates will be **(cooler/warmer)** than daytime temperatures in desert climates.

29. The warming of the atmosphere due to the presence of water vapor and other heat trapping gases is called the __________________________.

30. _____________ are airborne masses of tiny water droplets or ice crystals. Their formation depends upon the presence of particles in the atmosphere. List three types of particles in the atmosphere that provide surfaces for the condensation of water vapor. (1) _________ (2) __________ (3) ____________

31. The process responsible for the formation of clouds is ________________. The process that returns the moisture to Earth is ____________________. The amount of precipitation that falls depends upon these conditions: (1) __________ ____________________ (2) __________________________.

32. Cutting the tropical rain forest can affect weather pattern by reducing the amount of water vapor that enters the air through the process of ____________________.

33. Clouds act as a _______________ to reflect light energy and a ___________ that traps heat energy. The temperature of the planet is regulated by the amount of ______________________ and other heat-trapping gases in the atmosphere.

34. Life on planet Earth is possible because the atmosphere traps enough heat to allow water to exist as a _______________.

Seasonal Weather [103]

35. Because the shape of the Earth is nearly that of a ___________ all regions of the Earth do not receive the same amount of solar radiation. The region of the Earth that receives the most direct rays from the sun is the region near the **(equator/poles)**.

36. More solar radiation is absorbed at the equator because the sun's rays strike the equator at a(n) **(oblique/right)** angle.

37. The prevailing wind direction and ocean currents are determined by the Earth's **(rotation/shape)**.

38. The northern states experience summer and winter because the Earth **(rotates on its axis/rotates around the sun)** and **(is a sphere/is tilted on its axis)**.

    In the northern hemisphere, the days are shorter and temperatures colder when the **(Earth is farther from the sun/north pole is tilted away from the sun)**.

Name ______________________________ **Date** ________ **Class Period** ____

During the winter months, the angle that the sun strikes the northern hemisphere at an angle that is **(close to/much less than/much more than)** 90°.

Down to Earth [104]

39. Water gains and loses heat at a **(faster/slower)** rate than land. On a summer day, the temperature of land will be **(cooler/warmer)** than the temperature of a lake. After the sun sets the **(land/water)** will lose heat slowly and remain warmer for a longer period of time.

40. During the winter months, land near the ocean will be **(cooler/warmer)** than land that is not near a large body of water.

    During the summer months, land near the ocean will be **(cooler/warmer)** than land that is not near a large body of water.

41. The difference in temperature of water and land creates a difference in air pressure that causes the air to move and creates a breeze. At the seashore the breeze will flow from water to land during the **(day/night)** and from land to water during the **(day/night)**.

42. What two types of air masses must meet to create a storm? ________________ ________________________

43. As the air rises over a mountain, it is **(cooled/warmed)** and the water vapor **(condenses/evaporates)** and the result is some form of ________________.

44. Due to the prevailing wind direction in the northern hemisphere, the western side of mountain ranges will be **(drier/wetter)** than the eastern side.

45. Cities are usually **(cooler/warmer)** than the surrounding countryside due to the presence of many buildings and other structures that hold __________.

46. If there is no breeze, the flow of air is **(away from/toward)** the center of a city.

47. On windless days in cities where the air is filled with smoke from factories, the particles ____________ sunlight and the movement of air is **(faster/slower)** than air movement in cities without air pollution.

48. Normally cooler temperatures occur at **(lower/higher)** altitudes, but during an **inversion** the air near the Earth's surface is **(cooler/warmer)** than the layer of air above.

49. Inversions most often affect cities that are located in a ___________ where cool air becomes trapped by a layer of __________ air above. Movement of air depends upon energy from wind or __________________ to warm the Earth.

50. During which season<u>s</u> of the year do the longest lasting inversions occur? ______ and __________. Why do they occur more often at these times? __________ ____

51. Why are inversions sometimes dangerous? __________________________

Name ________________________ Date __________ Class Period ____

Unit II — 2.2 Is the Earth's Climate Changing?

Introduction [109]

1. Compare climate and weather. ______________________________
____
____

Global Warming [109]

2. Gases that trap heat and warm the planet are called ____________________.

3. The warming of the Earth caused by certain gases in the atmosphere is called the ____________________.

4. An increase in the planet's average temperature is called ____________________.

5. According to weather records, during the last 100 years the Earth's average temperature has increased about ____ degree(s) Fahrenheit or ____ degree Celsius.

   This temperature change may be due to ____________________ or it could be due to ____________________.

The Greenhouse Gases [110]

6. The most important atmospheric greenhouse gas is ____________________. The amount of this gas present in the atmosphere is **(is/is not)** changing.

7. Place a checkmark beside the statement that is true.
   [ ] Water vapor contributes to the greenhouse effect.
   [ ] Water vapor is responsible for global warming.

8. An increase in water vapor may have a warming or cooling effect on the average global temperature. Describe the type of clouds that would have a:
   cooling effect: ______________________________
   warming effect: ______________________________

9. The most important greenhouse gas that is added to the atmosphere by human activity is ____________________. Prior to the industrial revolution the level of this gas was about ________ ppm and today it is about ________ ppm.

10. Scientists have studied the relationship between atmospheric gases and air temperature by analyzing ______________. They have discovered that increases in temperature are accompanied by increases in these two gases: ______________ ______________ and ______________.

11. During the next decade scientists expect the carbon dioxide level to ____________. The major source of carbon dioxide is ______________________________.

Name ______________________________ Date __________ Class Period ____

12. Place the major human activities that produce carbon dioxide in order beginning with the activity that produces most of the carbon dioxide.
(1) ____________________ (2) ________________ (3) ________________

What additional human activity contributes to the increase in carbon dioxide?
____

13. What process is primarily responsible for the increase in the level of methane?
____

14. Classify sources of methane as:
Natural sources: ____
Human activities: ____

15. In addition to carbon dioxide and methane, other important greenhouse gases include: (1) ______________ (2) ______________ (3) ______________

The amounts of these gases in the atmosphere are very small compared to the level of carbon dioxide. Give one reason why are scientists concerned about each of these gases.
(1) ____
(2) ____
(3) ____

16. Briefly explain the process associated with each of the following that contributes to the level of nitrous oxide in the atmosphere:
Fertilizer: ____
Fossil fuels and vegetation: ____
Nylon ____

Turning the Thermostat Down [111]

17. What is the difference between the average surface temperature today and the average surface temperature that produced the Little Ice Age? ____________

18. What events that may have caused 1816 to be a "year without a summer."
(1) ____
(2) ____

19. Explain how eruption of a volcano can change the average surface temperature.
____

Living in a Warmer World [111]

20. Which of the following statements reflects the position of the World Resources Institute.
[ ] Greenhouse warming will be a very serious problem during the next 100 years.
[ ] It is unlikely that greenhouse warming will cause dramatic changes in climate during the next 100 years.

Name ______________________________ Date __________ Class Period ____

21. Fill in the blanks to show the possible effects of global warming.

Rise in sea level	• threaten people's ______________ • flood coastal ______________ habitat • intrusion of ______________ water into water supplies
Pattern of precipitation	• drier Great Plains will affect ______________ • threaten water ______________ in drier regions. • lower water levels in Great Lakes affects __________ • affect generation of ______________ power
Climate zone shifts	• plants may not be able to ______________ to changes • range of some plants and some ecosystems (including hardwood forests) will ______________ • National Parks may not provide the correct ______________ for wildlife • migratory animals may not be able to find __________
Agriculture	• increase in carbon dioxide may increase __________ of major crops • excessive heat may reduce ______________ of corn • high temperature may prevent development of _______ • problems with __________ and __________ increase
Weather predictions	• scientists _______ know effects on weather prediction • warmer sea surface ______________ will cause more ______________ and more ______________ storms especially along the coast.
Humans	• heat will intensify effects of ______________ • group most affected will be the ______________ • increase in _______ rate in large __________ and developing __________

Must We Act Now? [113]

22. Scientists are concerned about a change in temperature of only 1°C within a 10 year period. They feel that the ________ of change is too ___________ and organisms will not be able to ____________ . They are also concerned that the changes may be ______________ rather than gradual.

23. Computer models suggest that the temperature change may be moderated by the increased ____________ of water and the formation of ______________ .

24. Predictions made by computer models may not be completely accurate because of a lack of knowledge about the influence of ________ temperatures and __________.

25. Most scientists feel that we should **(act now to stop/wait until we know more about)** global warming.

26. Give two reasons to support immediate action to stop or slow global warming.
(1) ____
____
(2) ____
____

Name ______________________ Date ________ Class Period ____

What Can You Do? [113]

27. Reducing the threat of global warming requires reducing the ______________ of greenhouse gases can be done by increasing energy ________________ and using ______________ energy sources.

28. List actions you could take to help decrease the threat of global warming. For each action listed, explain how the action will reduce the threat.

Action	Effect
1.	
2.	
3.	
4.	
5.	

29. Justify the following measures taken to reduce global warming.
A. Switch from coal to oil or natural gas as a source of energy for the production of electricity: ______________________________
B. Build a nuclear power plant rather than an oil-burning power plant: __________
____
C. Cost of planting one million square miles of trees ($100 billion): __________
____

International Cooperation [114]

30. Describe the connection between global warming and each of the following:

Human population growth: ______________________________
____
____

Our lifestyle: ______________________________
____
____

Use of technology: ______________________________
____
____

31. Japan and some European countries are committed to stabilizing or reducing carbon dioxide levels. Do you think that the United States should be committed to a policy to reduce carbon dioxide emissions. _____ Explain your position.
____
____
____
____

Name ______________________________ Date ________ Class Period ____

Unit II — 3.1 The Air Around Us

Introduction [117]

1. A substance is a pollutant if it is present in an ______________ that is ________________ or ______________.

2. **(True/False)** Air pollution was not a problem before the early 1970s.

3. The skies of London were darkened by smoke from the burning of __________.

Processes the Pollute [117]

4. Identify the three processes that pollute the atmosphere.
(1) ______________ (2) ______________ (3) ________________

5. Identify the process which is responsible for adding substances to the air in each of the following activities.

(1) attrition (2) combustion (3) vaporization

____ A. a romantic evening with a candlelight dinner and a fire in the fireplace.
____ B. spray painting the Golden Gate Bridge.
____ C. setting the thermostat in your classroom at a higher temperature.
____ D. enjoying the aroma of your favorite perfume.
____ E. sanding a project in wood shop.
____ F. hanging your woolen sweaters outside to get rid of the mothball odor.
____ G. burning the trash.
____ H. production of electricity at a power plant that uses coal.
____ I. using any household cleaner that is in a pump bottle or aerosol can.

6. Which process is responsible for most odors? ____________________

7. List the three conditions which must exist in order for the only products of combustion to be carbon dioxide and water.
(1) ____
(2) ____
(3) ____

Nature vs. Man [118]

8. According to the quote from the Manufacturing Chemists Association, the major source of air pollution is **(man/nature)**.

9. According to the quote from the League of Women Voters, the major source of air pollution is **(human activities/nature)**.

10. What do you think is the major source of air pollution? ____________________

11. Why do you think the Manufacturing Chemists Association and the League of Women Voters had different views about themajor source of air pollution?
____
____

Name ______________________________ Date __________ Class Period ____

Nature Pollutes [119]

12. The two major air pollutants emitted by volcanic eruptions are ______________ and ______________________.

13. The source of air pollution which killed people and animals in the valley near Lake Nyos was **(human activities/nature)**.

14. The killer cloud contained deadly amounts of **(carbon dioxide/carbon monoxide)**. The cloud of gas remained in the valley because of **(an inversion /the gas was heavier than air)**.

Urban Pollution [120]

15. When all human activities are considered, the major process that produces most air pollution is ____________________.

16. The major pollutants in Mexico City are ______________, ______________, ______________, and ______________.

17. What evidence is there that these pollutants are harmful? ____________________
____

18. The major sources of Mexico City's air pollution are ______________ and an ______________.

19. The major pollutants in the city of Prague are ____________________ and ______________________________.

The air is also polluted with traces of ______________, ______________, and ______________.

20. The major source of the pollutants is ______________________________

21. What evidence exists that Prague's pollution reaches harmful levels? __________
____

22. The major air pollutants in urban areas in the United States are ______________, ____________________, and ____________________.

The biggest sources of air pollutants in urban areas is ____________________.

23. Has the government kept the promise made in the 1970 Clean Air Act? ________

Progress has been made in clearing the air.
What city met the ozone standard in 1990? ______________________

Smog caused **(fewer/more)** problems in Southern California in 1991.

Name ________________________________ Date __________ Class Period ____

Unit II — 4.1 Chemicals that Pollute the Air

Introduction [122]

1. Before entering a mine, miners sent canaries down the mine shafts to test the chemicals in the ________.

2. According to Roger Tory Peterson, birds are like litmus paper—they indicate the ______________ of the environment.

3. Pollutants often affect birds more quickly than humans because of the bird's small ____________ and high __________________________.

4. What do you think that the employees actions at the hotel in Kuwait?
____
____

5. What toxic chemicals used today are a by-product of developing agents for chemical warfare? ____________________________

6. The first synthetic pesticide that was used to protect soldiers against insects is known by these three letters __________.

7. Pesticides eliminate _____________ and ______________ that cause illness in animals and loss of ______________.

8. Pesticides may pollute the air during both ______________________ and ______________________. The processes responsible for air pollution are __________________________ and _________________________.

9. What evidence indicates that pesticides travel long distances and makes pesticide use a global environmental issue? ____________________________________
____
____

Volatile Organic Compounds [123]

10. What are the two characteristics of all Volatile Organic Compounds (VOCs)
(1) _________________________ (2) __________________________________

11. Identify natural sources and industries or businesses that release VOCs into the atmosphere.
Industries/Businesses: ____
Natural sources: ____

Identify the processes that put VOCs into the atmosphere: ____________________
____

12. Hydrocarbons contain the elements __________________ and _______________.
and they ______________ at normal temperatures. Are hydrocarbons VOCs? ____

Name ______________________________ **Date** ________ **Class Period** ____

13. Hydrocarbons that are fuels include __________, ___________and ___________. Compare these fuels. How do they differ? ______________________________
____

14. Gasoline is a mixture of hydrocarbons including ____________. The hydrocarbons in gasoline and other fuels pollute the air by the process of ________________ during ______________ and _______________ combustion.

15. Miners used the canary to detect the hydrocarbon _______________ in coal mines.

16. Today large amounts of VOCs are generated at __________________________ and _________________. The VOCs entering the air can be reduced by collecting the ________________ and burning it to produce _________________ or _________________________. If it is not collected, it becomes an air pollutant by the process of __________________.

17. List three consumer products that contain VOCs. __________________________
____

18. How is a car like a can of hair spray? ______________________________
____

By reducing VOCs entering the air, the state of California hopes to reduce the level of ground-level ________________.

Nitrogen Oxides [123]

19. The major **human activities** that produce nitrogen oxides involve the burning of ________________. The major sources of nitrogen oxides are _______________ _____________ and _____________.

Natural sources: Nitrogen oxides are formed by the action of ____________ in the soil. They are also formed naturally when nitrogen and oxygen in the atmosphere are exposed to the intense heat produced by ____________________, __________________, __________________, or ____________________.

20. Nitrogen oxide reacts with _____________ in the atmosphere to form a brownish red gas _______________________. In addition to the pollutant nitrogen oxide, pollutants classified as _________s must be present for this chemical reaction.

21. The heat and energy for the chemical reaction above are provided by ___________ and the term that describes the resulting pollution is ________________________.

22. An important pollutant in photochemical smog that is a form of oxygen containing three oxygen atoms is ___________. Since it is produced by chemical reactions in the atmosphere, it is a ______________ pollutant. Is nitrogen dioxide a secondary pollutant? (Hint: see question 20) _______

23. **Think about it!** List reasons why it is likely that Los Angeles often has higher levels of ozone than Chicago. (1) _______________________________, (2) __________________________, (3) ____________________________

Name ______________________ **Date** ________ **Class Period** ____

24. Explain why ozone in the lower atmosphere is considered to be a pollutant, but ozone in the upper atmosphere is not. ______________________
____
____

Carbon Oxides [124]

25. Carbon dioxide is always produced during combustion. Carbon monoxide is produced if there is not enough oxygen and combustion is ____________
Most carbon monoxide is produced by ______________________.
Thus the major source of carbon monoxide is **(industry/transportation)**

Other important sources are industry and the burning of ____________.
Smaller amounts of carbon monoxide are produced in __________ fires and underground ________ mines.

26. Carbon dioxide **(is/is not)** usually considered an air pollutant because it is not found in amounts that are ________. Carbon dioxide will be classified as a pollutant scientists find that it is harming the Earth's ________.

27. The country that produces the most carbon dioxide is the ____________.

Suspended Particulates [124]

28. **Suspended** ______________ include __________ particles or ________ small enough to ______________________.

29. Give examples of natural sources of suspended particulates and human activities that produce suspended particulates.
Natural sources: ______________________
Human activities: ______________________

30. Incomplete burning of fuels produces small carbon particles called ________ and small pieces of minerals called ____________.

Fly ash may contain toxic chemicals such as ________ and ________.They also serve as a vehicle for other ____________ to travel into the lungs.

31. Suspended particulates in diesel exhaust are in the form of solid particles (soot) and small droplets called __________. Other sources of aerosols are ____________ and ______________.

32. Suspended particulates from industries and the burning of fossil fuels may be the source of __________ found in remote lakes in Minnesota. Suspended particulates are the source of ________ and ____________ that are found in the polar ice caps.

Sulfur Dioxide [125]

33. In addition to hydrogen and carbon, coal and fuel oil often contain the element ________. When these fuels are burned, this element combines with __________ from the air to form ___________.

34. The process that produces most sulfur dioxide is **(combustion/vaporization)**. Most of the sulfur dioxide that enters the atmosphere is produced by **(electric power plants/industries)**

35. Sulfur dioxide (SO_2) combines with _______________ in the atmosphere to form sulfur trioxide (SO_3). Sulfur trioxide combines with ________________ in the atmosphere to form another pollutant ________________________ (H_2SO_4).

36. An aerosol consisting of small droplets of sulfuric acid creates a ____________ that scatters light and reduces ___________________ .

Indoor Air Pollution [125]

37. What are the symptoms of "sick building syndrome"? ______________________
____

38. What is the cause of "sick building syndrome"? ____________________________

39. Identify potential sources of pollutants that cause "sick building syndrome."
____
____

40. List the sources of each of the following indoor air pollutants.

Formaldehyde: ____
____

Nitrogen oxide: ____
____

Carbon monoxide: ____
____

Particulates: ____

VOCs: ____

Name ______________________ Date ________ Class Period ____

Unit II — 5.1 The Effects of Air Pollution

The Disasters [127]

1. The major source of the air pollution that caused the major air pollution disasters was **(transportation/the burning of coal)**.

2. Fog that is contaminated with smoke is called __________.

3. Weather conditions which lead to an air pollution disaster include **(stagnant air/strong winds)**.

4. Air pollution disasters often occur when there is a temperature ______________.

6. Weather conditions that caused the air pollution disasters occurred during the **(fall and winter/spring and summer)**.

7. Describe the geographic location of Donora, Pennsylvania.
____
Did the other air pollution disasters also occur in valleys? ________
How does this location contribute to the air pollution disasters? HINT: see section 2.1 Weather Patterns. ______________________________________
____

8. What groups of people were most affected by the air pollution disasters?
____
____

9. Immediately after the major air pollution disasters, the United States acted quickly to eliminate the problem. **(true/false)**.

In the Laboratory [128]

10. In a controlled experiment subjects are divided into _______ groups. There must be only ______ difference or ________________ between the groups.

11. Mice and other laboratory animals exposed to nitrogen oxide and auto exhaust developed lung **(cancer/infections)**.

12. ________________ in tobacco and cigarette smoke produce cancer in laboratory animals. Although laboratory animals will not voluntarily "smoke" as humans do, animals that are forced to breathe cigarette smoke develop lung cancer. Do you think that these studies can be applied to humans? ________ Why or why not?
____
____

The Human Population — What the Studies Show [128]

13. Why is it difficult to study the effects of air pollution in humans? ____________
____
____

Name ______________________ Date ________ Class Period ____

14. One variable that must be considered in an air pollution study is that some people smoke cigarettes while others do not. List other variables that may occur in a study of air pollution. ______________________________

____

15. Studies have shown that long-term exposure to air pollution may be unpleasant, but it is not in any way harmful. **(true/false)**

16. Studies have shown that your risk of respiratory problems is higher if you live in a section of the city with higher levels of air pollution. **(true/false)**
Give two studies that support your answer. ______________________________

____

17. Studies have shown that people living in cities with higher levels of air pollution are more likely to respiratory problems. **(true/false)**
Give two studies that support your answer. ______________________________

____

18. The Los Angeles, California, study showed a direct relationship between the level of air pollution and the number of people admitted to hospitals who were suffering from heart disease. **(true/false)**

19. Identify the variables which were studied in the St. Louis–Winnipeg study.

____

Do you think this was a good study? ________ Why or why not? ____________

____

____

20. Compare the St. Louis–Winnipeg study to the Longbeach–Glendora–Lancaster study. Is the design of one study better than the other? ____________________

____

____

What the Surgeon General Says [129]

21. Identify the increased risk to smokers for each of the following.
(1) lung cancer ________
(2) lung diseases (such as chronic bronchitis or emphysema) _______
(3) coronary heart disease _______

22. Identify two forms of cancer that are more common in smokers.
(1) ______________________ (2) ____________________

23. List four factors associated with smoking that affect the chance of the smoker developing cancer.
(1) ______________________________
(2) ______________________________
(3) ______________________________
(4) ______________________________

24. After a person stops smoking the chance of developing cancer decreases. **(true/false)**

Name ____ Date __________ Class Period ____

25. A pollutant causes more damage when present with another pollutant than it would cause if it were the only pollutant. The increase in damage is called the __________ ______________________

Cigarette smoke causes more damage in the presence of other air pollutants than if it is the only type of pollution. **(true/false)**

City smokers have **(a much greater cancer rate/the same cancer rate)** as rural smokers.

The increased damage is due to the ___________________________.

Diseases Associated with Air Pollution [130]

26. What two changes in the respiratory system bring about the chronic cough which is sometimes known as "smokers cough"?
(1) ______________________________ (2) ________________________________

What is this condition called? ______________________________________

27. What changes occur in people who develop an allergic reaction to certain pollutants?
(1) ______________________________ (2) ______________________________
(3) ______________________________

What is this condition called? ____________________________________

28. What disease results when the air sacs remain inflated and can't help push the air out of the air sacs? ____________________________________

29. Pollutants cause changes in the respiratory system. Which changes will encourage the growth of bacteria?(1) ____
(2) ____

30. Which changes listed in questions 26–28 will decrease the amount of air moving through the respiratory system? (1) ______________________________________
(2) ____
(3) ____
(4) ____

31. How do cancer cells differ from normal cells found in the lung?
(1) ____
(2) ____

Pollutants — How They Affect Humans [131]

32. Identify the two types of pollutants in photochemical smog and cigarette smoke.
(1) ______________________________ (2) ______________________________

33. Place a **C** beside those substances that are carcinogens and an **I** beside those substances that are respiratory irritants.

_____ asbestos	_____ chlorine	_____ radioactive substances
_____ ammonia	_____ formaldehyde	_____ sulfur dioxide
_____ arsenic	_____ nitrogen dioxide	_____ sulfuric acid
_____ benzene	_____ ozone	_____ vinyl chloride

Name ______________________________ Date ________ Class Period ____

34. The gas sulfur dioxide is only mildly irritating. But under certain conditions it is deadly. Describe the conditions that increase the danger.
(1) ____
(2) ____

35. Identify two air pollutants that increase the risk for people with asthma.
(1) ____________________ (2) ____________________

36. Children less than two years old whose parents smoke are more likely to be hospitalized with ____________________________________

37. Some pollutants enter the lungs as a gas. Other pollutants enter the lungs on the surface of particles. Some particles are trapped in the ____________ and swept out by ____________. Particles that are very **(large/small)** may be carried deep into the lungs.

38. Place a checkmark beside the statements that are true. Studies show that particulates:
[ ] increase respiratory infections. [ ] significantly increase death rates.

39. How does carbon monoxide interfere with the body's ability to get oxygen?
____
List three effects of moderate levels of carbon monoxide. ________________
____

40. If a woman smokes, do you think that she should quit smoking if she becomes pregnant ? _______ Justify your position. ____________________
____

When a woman becomes pregnant, do you think that other people living in the home should give up smoking or at least refrain from smoking in the house?
_______ Explain.____________________________________
____

41. What pollutant caused hatters to become "mad"? ______________

42. Mercury and lead affect the ______________ system and the ____________.

43. At low levels lead interferes with the action of vitamin ______ that is necessary for normal ______________________ in young children.

44. Predict the lowest level of lead that you would expect to find in the blood of a child who has these symptoms due to an accumulation of lead in the blood.

__________ red blood cell count is below normal
__________ vomiting and severe abdominal pains
__________ abnormal behavior and difficulty learning concepts
Note: These symptoms may be due to other causes.

45. Explain why children are more affected by lead and other air pollutants than adults?
____
____
____

Name ________________________________ Date _________ Class Period ____

Effects on Animals [132]

46. Which animals at the cattle show were most affected by the air pollution disaster in London? ______________________

47. Why should we be cautious about applying the results of an study of the effects of air pollution on laboratory animals to humans? ____________________________
____

48. Air pollutants may enter the bloodstream of animals from the air that they breathe. Another body system that is a pathway for air pollutants to enter the blood stream is the ____________________________

49. Identify two air pollutants that may cause animals to become lame.
(1) ________________ (2) ________________

Effects on Plants [133]

50. The death of plants downwind from copper smelters was due to ______________.

Describe the appearance of the leaves of a plant that is damaged by this pollutant.
____

Further away from the smelters the lower levels of sulfur dioxide may have reduced the growth of plants and decreased the ______________ of crops.

51. Using the chart, identify the crop in each pair (column) that would be most affected by low levels of sulfur dioxide pollution.

_____ lettuce	_____ corn	_____ marigolds
_____ tomatoes	_____ broccoli	_____ chrysanthemum

52. The damage from sulfur dioxide is greater if there is __________________

In the presence of sulfur dioxide this pollutant becomes more toxic. ____________
This a __________________ effect.

53. The EPA found that ____________ is the most damaging air pollutant. The level of this pollutant is costing __________ millions of dollars by reducing crop _______.

How does air pollution affect your choice of foods at the grocery store?
____

How does air pollution affect people in developing countries? ________________
____

54. You own land that you would like to use for growing a crop of pumpkins. The area is sometimes covered with a blanket of smog. Do you think a pumpkin crop will be successful? ___________________________.

55. Which variety of pine is most sensitive to smog? ________________

Name ______________________________ **Date** ________ **Class Period** ____

56. The hydrocarbon in automobile exhaust that interferes with the opening of some flowers is ____________________.

57. Why do you think the gingko tree is often planted along city streets? ____________ ____.

Property Damage [135]

58. What types of materials are damaged when sulfur dioxide and carbon dioxide mix with moisture to form acids? ____________________________________

59. What materials are damaged when ozone is present? ________________________ ____

60. Over time paper becomes brittle because ______________ in the air reacts with chemicals in the paper to produce ________________.

Haze — Spoiling the View [135]

61. On a clear day the ___________________ is reduced by a regional haze that is due to the presence of ______________________.

The main source of this pollutant is ______________________________________.

Vocabulary Review

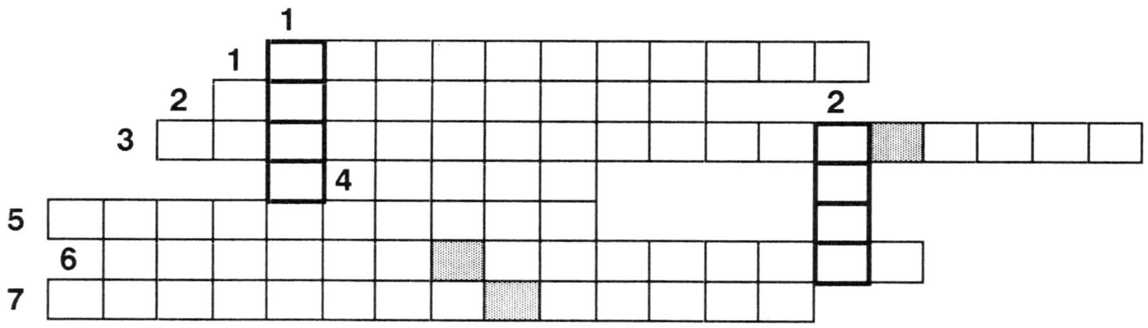

ACROSS:

1 the toxic effect of one pollutant is greater in the presence of another pollutant.
2 smokers often develop severe cases of this disease.
3 type of pollution that is created when the sunlight provides energy for reactions between certain chemicals in the atmosphere.
4 a heavy metal that causes some animals to become lame.
5 a substance that causes cancer.
6 a chemical that attaches to red blood cells more quickly than oxygen.
7 this gas reacts with oxygen to form the brownish red gas that is characteristic of photochemical smog.

DOWN:

1 fog that is contaminated with smoke.
2 a heavy metal that damages the nervous system.

Unit II — 5.2 The Blue Mountain — In the Shadow of a Smelter

The Franklin Ore [138]

1. Some—but not all—rocks are classified as an **ore**. To be classified as an ore a rock must ____________________.

2. The ore that became known as the Franklin ore contained a unique collection of minerals including __________, __________, & __________.

3. You discover an important deposit of copper. You stake a claim and hire workers to __________ the ore. [Environmental Impact #1]
You buy a coal mine and hire more miners. [Environmental Impact #2]
You need the abundant supply of fuel for __________ the ore.
[Environmental Impact #3] This technology is used to separate __________ from __________ like copper.

Zinc Oxide — The First Product [138]

4. Identify 5 <u>industries</u> that use zinc oxide as an additive in their products.
1.__________ 4.__________
2.__________ 5.__________
3.__________ 6.__________

5. List three products that you use that may contain zinc oxide.
1.__________ 3.__________
2.__________

Metal from the Franklin Ore [139]

6. Explain how zinc is used in the following processes.
Galvanizing __________
Fusion __________
Die casting __________

7. Make the connection between the following substances by writing two or three words on each connecting line.
BRASS __________ ALLOY __________ ZINC

8. **Think about it!** Could you have come to school today without zinc and/or zinc oxide? __________ Is zinc an essential element? __________

Death on the Blue Mountain [139]

9. Compare the Blue Mountain in 1900 with the Blue Mountain in 1970.
____
____
____
____

10. List positive and negative impacts associated with smelting of metals.
\+ __________ + __________
– __________ – __________

11. Identify the factors that led to the disappearance of the original forest ecosystem on the Blue Mountain. ____________________

12. Identify the process that usually produces a new forest following a forest fire.

____

A Student Studies the Mountain [140]

13. Explain why Dr. Jordan suspected that the NJZ smelters were the sources of pollutants affecting the mountain. ____________________

____

14. Identify the major pollutants produced during the smelting of the zinc ores.

____

15. Refer to the data in the chart "The Blue Mountain—a Population Census" and answer the questions below by inserting the symbol > "greater than" or < "less than."
    - The population density of trees (number per hectare) at Lehigh Gap is _____ the population density at the Control sites.
    - The biodiversity (number of species) of trees at the Control sites is ______ the biodiversity at the Lehigh Gap.
    - The number of species of seedlings at the Control sites was ______ the number of species of seedlings growing at the Lehigh Gap.
    - The distance between the smelters and the Control sites was _______ the distance between the smelters and the plots at Lehigh Gap.
    - Both the number and kinds of trees growing in plots near the smelter is _____ the number and kinds of trees growing in plots at Control sites.

Experiments in the Laboratory [141]

16. Soil samples collected near the smelter showed potentially toxic concentrations of these two metals. ____________________ and ____________________.

17. In the controlled experiment, Dr. Jordan compared soil samples taken __________ the smelters with soil samples taken __________ the smelters.

18. The greatest concentrations of zinc were in the **(organic matter/top soil)** taken from sites **(near/far from)** the smelters.

19. Compare the plants grown in the "near" soil to the plants grown in the "far" soil.

____

____

20. Identify the life processes that are inhibited or slowed by the pollutants in the soil near the smelters. ____________________

21. Dr. Jordan concluded that the pollutant preventing the regrowth of the forest was

____________________.

22. Dr. Jordan suggested that there may be a synergistic effect between zinc and cadmium. Explain how this affects the plants. ____________________

____

Life and Death on the Forest Floor [142]

23. Explain why the dead leaves and dead trees accumulate on the mountain, but in other forest ecosystems they disappear. ________________________________

____

24. Identify animal populations in the forest that appear to be suffering from the effects of the pollutants. ________________________________

25. Make the connections: Trace possible pathways that pollutants might take from the smelter to the living organisms.

26. Explain why woodlice and earthworms are more affected by the pollutants than other small organisms. ________________________________

27. Zinc is not equally toxic to all species. Explain why a species may be missing from the mountain even though it is not affected by zinc. ____________________

____

____

Effects on Wildlife and Farm Animals [142]

28. Explain why it is difficult for scientists to predict the effects of the smelting operations on song birds. ________________________________

____

29. Identify the heavy metals that may be affecting the white-tailed deer population and describe their effects. ________________________________

____

30. The farm animals most affected by the pollutants were **(young/old)** horses. Lesions found in joints were probably caused by ____________________. Osteoporosis and kidney disease are symptoms of ____________________

Effects on Human Health [143]

31. Identify the weakness in the zinc smelter—lung cancer study. ______________

____

32. List four groups of people who may be at the greatest risk because they are exposed to the highest levels of metals. ________________________________

____

33. The EPA is concerned about the presence of two metals in the soil near the smelters. They are concerned about cadmium because it damages the __________. The concern about lead is based on its affect on children's ________________.

34. In the comprehensive health study includes a Control Group. The only difference between the Control town (Jim Thorpe) and Palmerton is ____________________

____

HRD — Hazardous Dust to Metal Products [144]

35. Identify the source of EAF dust. ____

36. Describe the recycling operation at HRD. _______________________________

____

37. Identify the "positive impacts" associated with the recycling of EAF dust.

____

38. Identify the "negative impacts" associated with the recycling operation at HRD.

____

____

39. In Pennsylvania, it is illegal to store hazardous waste without a permit from the Department of Environmental Resources. If HRD applies for a permit to store the EAF dust, and it was your decision to issue or deny the permit, would you grant the company a permit? _________
Justify your position. ____

____

Recycling and the Environment [145]

40. Identify the "positive impacts" associated with the recycling of EAF dust.

____

____

41. Identify the "negative impacts" associated with the recycling operation at HRD.

____

____

HRD versus the EPA [146]

42. In Pennsylvania, it is illegal to store hazardous waste without a permit from the Department of Environmental Resources. If HRD applies for a permit to store the EAF dust, and it was your decision to issue or deny the permit, would you grant the company a permit? _________ Justify your position.

____

____

____

Unit II — 5.3 The EPA at the "Palmerton Zinc Site"

Introduction [149]

1. What environmental problem was the target of the Comprehensive Environmental Response, Compensation, and Liability Act of 1980 (CERCLA)? ____________
____

2. The Comprehensive Environmental Response, Compensation, and Liability Act of 1980 is know as CERCLA or the ____________________ law.

3. Tools are needed to do a job. The Superfund law gave the EPA two important tools that are needed to protect people and the environment from hazardous substances. These tools were: (1) ________________________________
(2) ____

4. In which of the following situations might you be identified by the EPA as a "responsible party" and force you to pay for the "clean-up?" _______________

a. You operate a business that stores a hazardous waste product at the site. The EPA discovers that the chemical is threatening the ground water at the site.
b. You are hauling a load of liquid fertilizer (ammonia) from the factory to the farm. There is an accident, and the ammonia leaks from the truck.
c. You purchase land and find that barrels of toxic chemicals have previously been buried on the property.
d. You are the owner of a company that produces batteries. A toxic waste is produced and sent to a treatment facility. The EPA discovers that the untreated waste was dumped into a stream.
e. You invested money in a business that you did not operate. You grew tired of your investment and sold your interest in the business. Several years later the EPA declares that the business is a hazardous waste site.

5. Identify the "responsible parties" who must clean-up or pay for clean up of the hazardous waste at the "Palmerton Zinc Site?" ______________________
____

Unit 1: Blue Mountain [149]

6. What type of hazardous substances threaten the environment. _______________

7. A "clean-up" plan may include removal of the hazardous substances from the site. Explain why removal is not an appropriate clean-up method on the mountain.
____

8. ____________________ is a type of solid waste collected from smoke stacks at coal-fired power plants.
____________________ is a solid waste that remains after sewage is processed at sewage treatment plants.

9. Describe how these two types of waste are being used to restore the forest ecosystem on the Blue Mountain.
____
____

10. What do you think about the "clean-up" plan? Will it restore the forest ecosystem? _________ Will is protect the stream? __________ Why do you think the plan will succeed or fail? ____

____

Unit 2: The Cinder Bank [150]

11. In what ways does the cinder bank threaten the environment? ________________

____

____

12. To decrease infiltration, the "clean-up" plan includes __________________ the cinder bank and then covering it with a cap made of ________. Both infiltration and runoff will be reduced if rainwater is absorbed by ________________ growing in ____________________ on top of the clay ___________.

13. The further protect the stream and the ground water, "clean-up" plan includes collecting and __________________ any water that runs off or flows through the cinder bank to remove ____________________ that are leached from the cinders.

14. The "clean-up" plan was put on hold because of the need to investigate a potential threat caused by _________ in the cinder bank.

Unit 3: The Valley [151]

15. Why did the EPA rule that "no immediate action was necessary" after reviewing the results of the soil tests taken in 1988? ________________________________

____

16. What is the EPA's top priority now? ____________________________________

____

17. What steps is the EPA taking to "clean-up" the valley? _____________________

____

____

____

____

18. The current and former owners of the New Jersey Zinc Company have been identified as "potentially responsible" for the contamination of the soil. A study is being conducted to identify the source(s) of the contamination. Identify other possible sources of the lead found in the house dust.

____

19. If you lived in Palmerton and had young children, how would you protect them from the potential danger of exposure to lead and cadmium?

____

____

Unit 4: Water [152]

20. Analysis of water samples taken from homes in Palmerton and Jim Thorpe revealed that the town with the largest number of samples with high levels of lead was ______________________. The EPA suspects that the lead in the water is coming from ____
21. The EPA concluded that the wells supplying drinking water for Palmerton's residents were **(safe/unsafe)**.

22. A resident is concerned that the town's water supply may become contaminated 5 or 10 years from now. Explain to this resident why the EPA considers contamination of the wells unlikely.

____
____
____

Life Goes On [152]

23. As the chairperson of the committee to oversee the landscaping of the ballpark, present a plan for the project. The goal of the plan is to provide an improved place for the area children to play sports.
THE PLAN:

____
____
____
____

JUSTIFICATION: Justify the expense to the Board Members of ZCA.

____
____
____
____

24. Do you think that the ballpark was dangerous before it was landscaped? ________ Why or why not? ____
____
Do you think it is safe now? ___________

The Clean Up Continues [153]

25. The "Palmerton Zinc Site" was listed as a Superfund hazardous waste site in 1982. Why is the "clean-up" taking so long? ______________________________
____

26. If you were a Palmerton resident, which citizen's group would you most likely support? _____________ Why would you choose this group? _______________
____

Name ______________________________ Date ________ Class Period ____

Unit II — 5.4 Disaster in Bhopal

Introduction [155]

1. The world's biggest industrial accident occurred at a plant that made **(batteries/missiles/pesticides)**. The plant was located in **(China/Great Britain/India/United States)**.

2. The chain of events leading to the release of the poisonous gas began when __________ entered a tank containing methyl isocyanate (MIC). The chemical reaction caused pressure in the tank to **(decrease/increase)**.

3. The gas escaped because **(the safety system failed/there was no safety system)**.

The Human Cost [155]

4. Some of the MIC (methyl isocyanate) released broke down into __________ ______________ a chemical used in the ________________ during W.W.II.

5. An antidote for cyanide poisoning was not used because:
[ ] it was not available.
[ ] doctors were not aware that the chemical MIC would form cyanide.

6. Those that didn't die from exposure to the cyanide suffered damage to the _______.

7. Which of the following statements best describes how MIC poisoning affected women.
[ ] the number of miscarriages in the months following the accident was double the expected number.
[ ] many young women who survived the accident are not able to have children.
[ ] women who were pregnant at the time of the accident gave birth to babies with birth defects.

Safety Tests [155]

8. A study conducted by the National Academy of Sciences concluded that information on health effects of most chemicals sold in the U.S.:
[ ] is adequate
[ ] is not adequate

9. Testing a new chemical requires several **(weeks/months/years)**.

10. Tests to determine the poisonous effects of a chemical usually require:
[ ] computer programs to predict chemical reactions.
[ ] hundreds of laboratory animals as test subjects.
[ ] a large group of volunteers as human test subjects.

11. The **(federal government/manufacturer)** is responsible for conducting safety tests on new products.

12. How are the test results used? ______________________________________
____

Name ______________________________ Date __________ Class Period ____

13. Are most chemicals on the market dangerous? ______________________________

14. Testimony at a Senate hearing revealed the manufacturers have sometimes submitted fake data. **(true/false)**

15. The group of laboratory animals that receives the chemical being tested is the **(control group/experimental group)**.

16. What is the purpose of the control group? ______________________________
____

17. A group of rats is fed the wrong diet for one month during the testing of a chemical. What should the scientists in charge of the study do with the data collected from this group of rats? ______________________________
____

18. ____________________ is the study of the effects of poisonous. Give two reasons why the use of data from animal studies to calculate the risks a chemical presents to humans is often questioned?
(1) ______________________________
(2) ______________________________

19. What are the advantages of the tissue culture technique?
(1) ______________________________ (2) ______________________________

With these advantages, why are animal studies still used? ____________________
____
____

20. Which of the following statements is true?
[ ] Union Carbide faked the safety tests for methyl isocyanate.
[ ] Union Carbide manufactured methyl isocyanate in India to avoid safety tests.
[ ] Union Carbide was required to submit safety test results because MIC is also manufactured in the United States.

21. In studies of rats exposed to large doses of MIC the system that was affected was the **(immune/reproductive/respiratory)** system.

The Question of Safety [157]

22. After reading the information in this section, do you think a disaster like Bhopal could occur in the U.S. ? ____________ Give reasons to support your answer.
____
____
____

23. What steps do you think should be taken to try to prevent another Bhopal disaster?
____
____
____
____

Name ______________________________ Date __________ Class Period ____

Preventing Accidents [158]

24. Who is responsible for safety at a chemical plant? ____________________

    List three steps that plant managers should take to reduce the risk of accidents at chemical plants?
    (1) ____________________
    (2) ____________________

25. What federal agency is responsible for the safety of workers at chemical plants and other industries? ____________________

26. List two requirements that chemical plants must meet to reduce the risk to the workers.
    (1) ______________ (2) ______________

27. How many of the 20 chemical plants located in or near your community are inspected each year? __________
    How often do you think a chemical plant should be inspected? __________
    Are you willing to pay more taxes so that more federal inspectors can be hired? __________ If not, how do you suggest the salary of federal inspectors be paid? ____________________

28. Good safety measures during transport can **(guarantee that no release will occur/reduce the risk of the release)** of a hazardous chemical.
    Identify the precautions taken by Dupont that may have prevented an accidental release of MIC from a shipment to New York.

    ____

    Identify one precaution taken by the state of Pennsylvania that may reduce deaths and injuries from accidental releases of hazardous chemicals.

    ____

29. Identify the law that requires communities and companies to work together to reduce the risk from chemical releases?

    ____

30. List the three requirements that the "Right-to-Know" Act places on industries.
    (1) ____________________
    (2) ____________________
    (3) ____________________

31. What does the "Right-to-Know" Act require of the community? ____________________

32. What is the best way to reduce the risk of an accident involving a toxic chemical?

    ____

    What if there are no alternatives to toxic chemicals? ____________________

Name ______________________________ Date ________ Class Period ____

Unit II — 5.5 Forecast: Acid Deposition (Acid Rain)

Acid Rain: A Political Issue [161]

1. Why is acid rain a source of political dispute between countries? ________________
____
____

2. The term acid rain was first used in a speech before the United Nations in the year ________, but it was not until the year ________ that an official of the U.S. government officially acknowledged that acid rain is an environmental problem.

Acid Rain: Defined [162]

3. Identify the term that describes substances with the following pH numbers as acidic, basic, or neutral.
3 ______________ 8 ______________ 6 ______________
7 ______________ 4 ______________ 12 ______________

4. A substance with a pH of 4 is 10 times more acidic than a substance with a pH of ____ and 100 times more acidic than a substance with a pH of ____ .

5. Natural rainfall—without any pollutants—is acidic because it contains **(carbonic acid/nitric acid/sulfuric acid)**.

6. The pH of natural rainfall is usually generally between ________ and ________.

7. Explain why scientists prefer the term "acid deposition" rather than "acid rain."
____
____

8. Identify the wet forms of "acid rain." ________________________

9. Identify the dry forms of "acid rain." ________________________

10. Indicate whether the pH of rain is raised or lowered by each of the following events:
Dust blown from desert soils **(lowers/raises)** the pH of rain.
Lightning during a thunderstorm **(lowers/raises)** the pH of rain.
Sulfur oxide from hot springs **(lowers/raises)** the pH of rain.
Eruption of a volcano **(lowers/raises)** the pH of rain.
Decaying matter **(lowers/raises)** the pH of rain.

Human Activities and Acid Rain [163]

11. What is the biggest single source of chemicals that form acids when they are released in the atmosphere. ________________________

12. The two gases that form the major pollutants in acid rain are ________________ and ____________________.

Name ______________________ Date __________ Class Period ____

13. Identify the major source of:
sulfur dioxide in Canada: ______________________
sulfur dioxide in the United States: ______________________
nitrogen dioxide in Canada and the United States: ______________________
acid rain on the west coast: ______________________
acid rain in the Northeast: ______________________

14. The state that has the most acidic average rainfall is **(California/New Jersey/ Pennsylvania/West Virginia)**.
The average pH of the rainfall is ______________.

The Midwest-Northeast Controversy [164]

15. What method used to control pollution, during the 1960s and 1970s, actually led to increased levels of acids in the rain? ______________________.

16. How can it be proven that a chemical that lands in New Jersey or Pennsylvania actually came from a certain smoke stack in Ohio? ______________________
____

17. Give two types of chemicals that are used as tracers. (1) ______________________
____
(2) ______________________
____

18. The purpose of the experiment named Captex was: ______________________
____

19. What can experiments like Captex tell us about acid rain? ______________________
____

Acid Rain: Is It Deadly? [165]

20. A laboratory technician analyzes the blood of 3 patients. The pH readings were:
Patient A = 7.30 Patient B = 7.40 Patient C = 7.70.
Which patient had a normal blood pH reading? ________________
If the pH of the body cells is not correct the nervous system can not function properly. By the time the technician completes the tests one of these patients will be dead. Which one? ________________

21. The main concern about acid rain is damage to **(aquatic/forest)** ecosystems.

22. Which of these species of trout is the most sensitive to acid rain? **(brook trout/brown trout/rainbow trout)**

23. A biologist tests the pH of three lakes. The pH readings were: Lake A 5.0, Lake B 6.5, and Lake C 4.0. Which of these lakes probably had no fish? ________

24. Many species of fish will not be found in water with a pH below 6 because they cannot ____________________.

25. Lakes and streams often more acidic in **(spring/summer)**. During this time of year lakes and streams receive acid from ____________________.

Name ______________________________ Date ________ Class Period ____

26. Fish often die because of the high levels of acid in water during the early spring. This condition is called ____________________. This condition is a greater problem in **(Canada/United States)**.

27. Most spotted salamander eggs do not develop properly if the pH is below ______.

28. Researchers have found that water with a low pH usually contains **(lower/higher)** levels of minerals than water with a high pH. The ________ in the water dissolves the minerals from ________________.

29. The process in which minerals are dissolved and carried away by water is called ____________________.

30. The mineral that prevents the normal exchange of oxygen and carbon dioxide across the gills of fish is ________________. When too much of this mineral is present the fish produces too much ____________________ and the fish suffocates.

31. Fish become deformed because aluminum interferes with the absorption of ________________, and they die from spastic muscle contraction because it interferes with the absorption of __________________.

32. Species which can survive in water with a low pH are said to be ____________ ________________.

33. Identify the following insects as acid tolerant (T) or acid sensitive (S) insects.
____ **mayfly** ____ **stonefly** ____ **water boatman** ____ **whirligig beetle**

34. If you examine a lake and find only a few species of mosses and algae and only a few species of aquatic invertebrates, what pH might you expect to find? ________

35. Which statement best describes an acid lake?
[ ] An acid lake is a lake with only a few species of organisms, but large populations of each species.
[ ] An acid lake is a lake with many different species of organisms, but smaller populations of each species.

36. Layers of leaves are often found lying on the bottom of an acid lake. This same situation occurs in bogs. The leaves do not decay because there is a lack of __________________.

The Buffer Factor [167]

37. Why do some lakes become acidic while others do not?______________________

38. A ____________________ is a substance that prevents large changes in pH. It ________________ the acidic rain.

39. What is the source of buffers that are found in a lake? ____________________
____

40. Water in a lake that contains mineral that act as buffers is described as __________ __________________

Name ________________________________ Date _________ Class Period ____

41. Which lake will be most affected by acid rain—one with hard water or one with soft water? ______________

42. Which lake is more likely to become acidic—the lake in a watershed that contains **(granite/limestone)**.

43. The lake that contains the most buffers is a lake in a watershed that contains **(granite/limestone/sandstone/shale)**.

44. The major type of rock found in the watershed of the Great Lakes is limestone, therefore the Great Lakes contain **(hard/soft)** water.

45. Cellar Pond, in the Adirondack Mountains, is an example of a lake that is sensitive to acid rain. It has a pH of 3.8. What type of rock would you expect to find in the watershed surrounding Cellar Pond? ______________

46. According to the Pennsylvania Fish Commission, the greatest threat to future fishing in Pennsylvania streams is ______________.

Can the Ecosystem be Saved? [168]

47. In an attempt to save some of their lakes from acid rain, New York and Sweden are pouring ______________ into some of their lakes.

48. Scientists think that adding limestone to lakes:
[ ] is a good alternative to expensive pollution controls which will reduce acid rain.
[ ] may help save lakes until pollution controls reduce acid rain.

49. In some lakes the limestone has caused population explosions of certain species of __________.

Impacts on Field Crops [169]

50. **(Most/Few)** field crops seem to be affected by acid rain. Give a possible reason why some plants grow better when sprayed with acid rain ______________
____.

51. Most acids entering the soil where corn is grown come from **(acid rain/ fertilizers applied by the farmer)**.

52. Farmers prevent the soil in their fields from becoming acid by adding ______________.

53. Acid rain could damage the __________ of some plants and interfere with the process of ____________ reducing the __________ of some plants. Scientists **(have/have not)** found evidence that acid rain is damaging pastures and field crops.

54. Scientists **(do/do not)** think that our food supply is threatened by acid rain.

Name ______________________________ Date ________ Class Period ____

Impacts on Forests [169]

55. The condition that describes a forest showing signs of stress is ____________ ____________. Describe the appearance of foliage in a forest that is showing signs of decline. ____
____

56. Place a checkmark beside each of the following types of trees that are showing signs of damage on Camel's Hump:

____ young seedlings ____ old red spruce trees
____ fir trees ____ beech trees

57. The amount of organic matter is in the soil of the forests on Camel's Hump is **(decreasing/increasing)**. This shows that:
[ ] something is wrong with the forest ecosystem.
[ ] the forest is a healthy ecosystem.

58. Explain how acid rain may be changing the minerals in the soil.
____
____

59. Trees may be dying because of increased concentrations of the metal __________. Increased levels of this metal may be inhibiting the uptake of ______________. It may also be killing ____________________ in the soil.

60. What evidence indicates that the growth rate of red spruce at high elevations on Camels Hump is declining? ____

61. Winter damage in red spruce at high elevations is increased by exposure to acidic ______________. ________ growth of ponderosa pine and Douglas fir is affected by exposure to acid __________ and ____________________ (gas).

62. Research shows that there is a direct link between acid rain and:
[ ] widespread forest damage
[ ] changes in soil chemistry

63. According to the U.S. Forest Service, changes in soil chemistry will have a negative effect upon forests. **(true/false)**

64. According to scientists high levels of ______________ are more damaging to forests than acid rain.

65. List factors, other than acid rain, that may be causing trees to die. ____________
____

The Statue of Liberty [170]

66. Acid rain is often blamed for damage to structures including the Statue of Liberty. How much of the damage is actually due to acid rain? ____________________
____

Name ______________________ Date ________ Class Period ____

67. Air pollution, including acid rain, is everyone's problem. Explain this statement by describing how air pollution is costing you money. ______________________
____

Can Acid Rain Harm People? [171]

68. Even when it is not raining, sulfur gases in the atmosphere form "acid rain" or acid ______________ when they combine with moisture in your ____________.

69. Sulfur gases that form "acid rain" or acid aerosols in the lung may be responsible for the deaths of people suffering from certain respiratory diseases such as ____________________, ______________ or ________________.

70. Acid rain sometimes causes high levels of metals, including ____________ and ______________ in the drinking water.

71. High levels of lead in the drinking water can affect ____________ and ______________ development of the ____________ in children.

72. What should you do to protect your family if corrosion of the plumbing is responsible for high levels of lead and copper in water supplies? ______________
____

The True Cost [171]

73. The **true cost** of a product includes the cost of ______________________
____

74. Do you think you should be able to buy products without paying the "true cost"? ____________ Would you be willing to pay $20 more each year for your electric bill if it would reduce the amount of acid in the rain? __________

Getting On With It [172]

75. What was the conclusion of the National Acid Precipitation Assessment Program study ? ______________________________________
____

76. One of the goals of the 1990 Clean Air Act Amendments is a major reduction in ____________________ emissions. Is this also a goal of clean air regulation in Canada? ____________

77. The goal of the Canada–U.S. Air Quality Accord is reduction of ______________ ____________ emissions from stationary sources such as ________________ and ____________________.

The major source of nitrogen oxides in both the United States and Canada is ____________________. Reduction of nitrogen oxides from this source **(is/is not)** a goal of the Accord signed in 1991.

Name ______________________________ Date ________ Class Period ____

Vocabulary Review

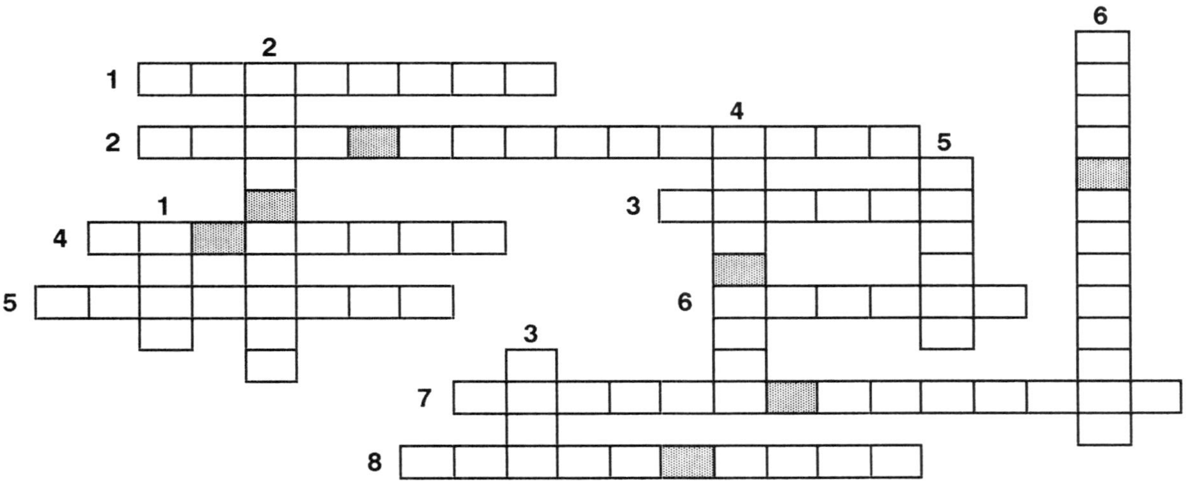

ACROSS

1 minerals are dissolved from rocks and carried away by water.
2 any form of moisture with a pH lower than natural rainfall.
3 a substance that, when placed in water, prevents large changes in the pH.
4 a range of numbers that is used to measure the acidity of a substance.
5 acid that is normally found in unpolluted rain.
6 an experiment that was designed to trace the path of a dangerous chemical explosion.
7 many trees show signs of stress including premature yellowing or loss of foliage.
8 the drainage area around a lake.

DOWN

1 water that contains high levels of minerals.
2 a sudden increase in acid from melting snow creates this condition that sometimes causes the death of entire populations of fish.
3 water containing low concentrations of dissolved minerals.
4 the price of a product including the technology that prevent pollution of our environment.
5 chemical added to emissions from a smoke stack in order to follow the path of pollution.

Unit II — 6.1 Clearing the Air — Stationary Sources

Introduction [175]

1. Pollutants from power plants, industries and vehicles damage the ______________ and affect human ______________.

2. ________________________ is an important source of lead in Lake Superior, nitrogen in the Chesapeake Bay and toxic chemicals in the Arctic ecosystem.

3. The air is cleaned by these natural scrubbers: ______________ and __________.

Air Quality Standards [175]

4. The federal agency that was created to control pollution is the ________________ ____________________. The agency was established in __________.

5. The EPA must set and enforce ______________________ the maximum amount of a pollutant allowed.

   ____________________ are gases and particles entering the air from smoke stacks, chimneys and exhaust pipes.
   An ______________ ______________ is the maximum amount of a pollutant that is allowed to enter the air (from a smoke stack, chimney or exhaust pipe).

The Air You Breathe [175]

6. The air outdoors or the air you breathe is called ________________ air.

7. **National Ambient Air Quality Standards (NAAQS)** were set for six major air pollutants in 1971. These pollutants are:
   carbon ____________ ____________ **dioxide** __________ **oxides**
   ground-level __________ ______________ matter __________

8. Pollution standards were set to protect the ______________ of the people. According to the EPA, in 1991 ______ of ten people lived in area that did not meet one of these standards.
   96 cities did not meet the NAAQ standard for ____________________
   72 cities did not meet the NAAQ standard for ____________________
   41 cities did not meet the NAAQ standard for ____________________

9. The air you breathe is called ______________ air while the gases coming from an exhaust pipe are called ______________.

10. Why do **emission standards** allow higher levels of pollutants than the **ambient air quality standards**? ____

    Why are cities finding it difficult to meet ambient air quality standards? ________
    ____
    ____

Setting Air Quality Standards [176]

11. Before setting standards for pollution control, the EPA must listen to information provided by two groups. These two groups are the ______________ (who present evidence of the effects caused by the pollutants) and the ______________ (who will be affected by the pollutants and/or the cost of preventing pollution).

12. The standards must protect ________________ and the ____________________.

    Standards must reflect the ________________ and ______________ of the public.
    The EPA must weigh the ___________ against the _____________.

13. Why do you think that EPA decisions are sometimes controversial? ____________
    ____
    ____

14. Pollution is harmful to people and the environment. Why does the EPA allow any pollutants to enter the environment? __________________________________
    ____
    ____

Controlling Air Pollution — Tall Stacks [177]

15. Why aren't "tall stacks" the answer to air pollution control? ___________________
    ____

16. Compare the amount of sulfur dioxide released by the nickel and copper smelting plant in Sudbury, Ontario, in 1970 with the amount released in 1987, and explain the change. ____
    ____

17. Identify two environmental groups that help protect the environment by taking polluters to court. ____
    ____

    Why did an environmental group take the EPA to court in 1983? ______________
    ____
    ____

18. The air is cleaner since the Clean Air Act was passed in 1970. Indicate direction of change (**I**) increase or (**D**) decrease for the following air pollutants.
    ____ sulfur dioxide ____ suspended particulate matter ____ nitrogen oxides

19. What is the goal of the 1990 Clean Air Act amendments? ____________________
    ____

The 1990 Clean Air Act Amendments [177]

20. Due to the "grandfather clause" in the 1977 amendments to the Clean Air Act, pollution controls were required on **(all power plants/only new power plants)**.
    How did the 1990 amendments affect power plants protected by the "grandfather clause"? ____
    ____

21. The 1990 amendments to the Clean Air Act require a reduction in the amount of ____________________ emissions.

22. The **emissions cap** sets a maximum amount of ________________________ allowed in the emissions from **(all power plants/new power plants)**.

23. Electric power companies are given "allowances" by the EPA. Each allowance gives the company the "____________ **to** ______________". One allowance permits the company to produce __________ ton of sulfur dioxide each year.

24. Before an electric power company builds a new power plant, the company must buy ____________________ for each ton of sulfur dioxide that the new plant will produce.

25. According to the EPA, removing the sulfur dioxide from the air will cost ______ per person or $25 billion each year. **Think about it!** Is cleaner air worth the cost? ________ What criteria would you use to decide? ____________________________
____
____

Why Burn Coal? [178]

26. Which contains more sulfur—soft or hard coal? ______________________

27. If sulfur is a dangerous pollutant, why don't we simply switch to hard coal and other fuels that are low in sulfur? ____
____

28. In addition to removing sulfur and reducing air pollution, what are two advantages of clean coal technology?
(1) ____
(2) ____

Clean Coal Technology — Cleaning Coal [179]

29. ____________ removes sulfur and other impurities from coal prior to combustion. Give advantages and disadvantages of this method of cleaning coal.
Advantage: ____
Disadvantage: ____

30. Scientist think that one day biotechnology techniques may produce a ___________ that will be able to break down organic compounds containing sulfur in crushed coal.

Clean Coal Technology — FBC [179]

31. **FBC** or ___________________________________ boilers burn ______________ coal and other low-grade fuels such as tires.

32. Pellets of ________________ and ______________ are injected into an FBC boiler along with ___________.

Calcium and sulfur react to form gypsum which is removed as ______________ or enters the smoke stack as ______________. Before the emissions leave the smoke stack the fly ash must be removed by ____________________________ or ____________________________..

33. FBC boilers:
(1) produce **(less/more) nitrogen oxides** due to **(higher/lower)** temperature.
(2) **(do/do not)** require scrubbers to remove **sulfur dioxide**.
(3) use crushed limestone which is **(cheaper/more expensive)** than powered limestone.
(4) produce **(less/more)** solid waste than other boilers.
(5) require **(less/more)** space than other boilers.

Clean Coal Technology — Wet Scrubbers [180]

34. A catalyst decreases the time it normally takes for sulfur dioxide (SO_2) to combine with ______________ and form sulfur ________________ (SO_3).

35. The sulfur trioxide (SO_3) is separated from other gases when it combines with __________ in the scrubber to form ____________________

$$SO_3 + H_2O \text{ -------->} \_\_\_\_\_\_\_\_\_\_\_\_\_\_$$

36. Sulfuric acid is used in the process of making many man-made products. Thus industries consume large amounts of sulfuric acid. Still coal-burning industries and power plants are unwilling to install catalytic converters and scrubbers unless required by law. Why? ____
____
____

37. Wet scrubbers are expensive because they require large amounts of ___________ to operate.

Removing Particles — Cyclones [180]

36. Cyclones are **(less expensive/more expensive)** than other methods for removing particles, but they are only effective in removing the ___________ particles.

37. _________________ must be used with cyclones to remove the smaller particles.

Removing Particles — Electrostatic Precipitators [181]

38. Before particles can be removed with this method, they must be given a strong ______________________. The charged particles are attracted to metal plates that have **(the same/an opposite)** charge.

39. The electrostatic precipitators remove ________ percent of the particles. This method is **(less efficient/more efficient)** than the cyclone.

40. This method is **(less/more)** expensive to operate than wet scrubbers.

Removing Particles — Filters [181]

41. Filters are very **(efficient/inefficient)** but they are **(expensive/inexpensive)** to install.

Removing Particles — Wood Stoves [181]

42. Particles entering the air from wood stoves can be reduced by using ________-__________ stoves, performing routine __________________, and burning ________ wood.

43. Following the steps listed above will reduce the particles entering the air by only _______ percent. This makes is necessary to prohibit the burning of wood in many areas when there is a ________________ weather pattern.

Removing Other Gases [181]

44. Nitrogen oxides contribute to the formation of these two types of air pollution:
ground-level __________ and **acid** ________.

45. What condition must exist for nitrogen and oxygen gases to form nitrogen oxides?
____

46. Two methods that reduce the nitrogen oxides in the emissions are regulating the flow of _________ and lowering the _______________________.

47. What two types of pollutants are reduced by recirculating flue gases?
(1) __________________________ (2) ________________________

48. Some of the nitrogen oxides polluting the atmosphere come from the oil and gas furnaces in our homes. Since technology is available to reduce nitrogen oxide emissions from furnaces, why will it take many years before the amount of nitrogen oxides from this source is reduced?
____
____

49. Reducing the ground-level ozone also depends upon reducing the level of VOCs or ____________ ____________ ____________ that are released by industries that use ____________, ____________, and ____________ processes.

Environmental Planning [182]

50. One way to reduce air pollution from electrical power plants is to reduce the need for electricity. The amount of electricity used at the Denver International Airport, is reduced by taking advantage of ___________ lighting and installing __________ ____________ lights.

51. Another way to reduce air pollution is to use ____________ energy sources or alternative fuels.

Name ____ Date __________ Class Period ____

Unit II — 6.2 Clearing the Air — Automobiles and Other Vehicles

Where it All Begins [184]

1. For each of the statements below, underline the choice that best completes the statement.
A. Carbon monoxide and hydrocarbons are waste products that are formed during **(complete/incomplete)** combustion.
B. In the **(external/internal)** combustion engine the fuel burns in a closed cylinder.
C. The fuels burns outside the engine in an **(external/internal)** combustion engine.
D. The steam engine is an example of an **(external/internal)** combustion engine.
E. The **(external/internal)** combustion engine used coal or wood as fuel.

2. Listed below are descriptions of actions that occur in an internal combustion engine. Identify the number of the stroke in which the action occurs.

____ The gas-air mixture is compressed.
____ Wastes are pushed out of the cylinder.
____ The gas-air mixture is drawn into the cylinder.
____ The spark plug ignites the gas-air mixture.

3. Identify the position of the piston (up or down) in each stroke of the four-stroke cycle.
(1) gas and air drawn in ________ (2) ______ gas and air mixture compressed
(3) gas and air mixture ignited ________ (4) wastes ejected from cylinder ______

4. The valves must be closed during strokes ______ and _____. This stroke is the power stroke _____.

Too Many Cars, Too Many Miles [184]

5. During the last twenty years, the number of vehicle miles traveled has **(decreased/increased)**.

If the rate of change is the same during the next twenty years, the number of miles traveled in the year 2010 will be ______________ miles.

If the percentage of total miles traveled doesn't change, how many of these miles will be traveled by car? ______________ miles.

6. Vehicles are the major source of three air pollutants in urban areas. The three air pollutants are _________________, ______________________, and ________ _____________.

The Government Steps In [185]

7. The state that first required pollution control on automobiles was ______________. Cars were required to meet standards for ____________________ and _________ _________________.

Name ______________________________ **Date** ________ **Class Period** ____

8. The Clean Air Act passed in 1970 set standards for the three major pollutants in auto exhaust. Identify the pollutant in each pair that is regulated by emission standards. **(carbon dioxide/carbon monoxide) (nitrogen oxides/sulfur oxides) (hydrocarbons/ozone)**

9. Identify the party responsible for each of the following.
   A. auto maker **B. Environmental Protection Agency**

   ____ Sets standards for pollution control on automobiles.
   ____ Makes decisions about the technology to meet the standards.
   ____ Certifies that cars meet the standards.

10. The major improvement in the quality of our air during the last 20 years is due to pollution controls placed on ______________________________.

The Catalytic Converter [185]

11. The catalyst in the catalytic converter is ____________________.

12. The speed of the chemical reactions listed below is increased by the action of catalysts in the catalytic converter. Complete the chemical reactions to show the products that are found in the vehicle's exhaust.

    Carbon monoxide + oxygen ----> ______________________________
    Hydrocarbons (H-C) + oxygen ----> ____________________ + ____________
    Nitrogen oxides ----> ____________________ + ______________

13. Which type of gasoline should be used in cars with catalytic converters? **(leaded/unleaded)** Why? ____

    ____

Get the Lead Out [186]

14. Why did refineries add lead to gasoline? Adding lead was the **(best/cheapest)** to produce **(high/low)** octane gasoline from **(high/low)** quality crude oil.

15. The high octane fuel prevents engine ______________ and ______________ the valves.

16. Give two reasons why the EPA made the use of leaded gasoline illegal.
    (1) ____
    (2) ____

17. Explain why the EPA did not immediately ban the use of lead.

    ____

18. Where is leaded gasoline still used today? ______________________________

    ____

Think about it! How does the use of lead in these countries affect you?

____

____

Name ______________________ **Date** ________ **Class Period** ____

Redesigning the Engine [186]

19. Identify two engines that produce less pollution than the original internal combustion engine. (1) ______________________
(2) ______________________

20. What are the advantages of the diesel engine over the original internal combustion engine? ______________________
____

21. What is the advantage of a gasoline engine over a diesel engine (without pollution controls)? ______________________

22. Less nitrogen oxides are produced in the main chamber of the stratified internal combustion engine because the ______________ is reduced.

23. If the temperature is too cool or the time the fuel mixture burns in the main chamber of the stratified internal combustion engine is too short, there is an increase in the amount of ______________ and ______________ produced.

Is It Time for a Tune-up? [187]

24. In addition to using the correct type of gasoline, what should car owners do to assure that their cars are not big polluters? ______________________
____

25. Match the engine part with its function (job).

(1) Air Filter
(2) Carburetor
(3) Cylinder
(4) Ignition System
(5) PCV System
(6) Spark Plugs

__ A. Cleans the air.
__ B. Contains pistons that compress fuel/air mixture
__ C. Recirculates vapors
__ D. Controls amount of air mixed with gasoline
__ E. Ignites fuel/air mixture in the cylinder
__ F. Causes spark plugs to fire

26. Match the major pollutant which will increase in each of the following situations.

(1) increased carbon monoxide
(2) increased hydrocarbons
(3) both of these

__ A. dirty or worn spark plugs
__ B. ignition system isn't working properly
__ C. PCV system clogged or air filter dirty
__ D. carburetor allows too much air to mix with gasoline.
__ E. leaks between cylinder and piston

Name ______________________________ Date ________ Class Period ____

Can We Meet the Standards? [187]

27. Identify the two changes significantly reduced pollution from vehicles?
(1) ______________________________
(2) ______________________________

28. The following are requirements for urban areas that do not meet the NAAQ standards.
Place a checkmark in the O_3 column to indicate actions that states must take in areas that do not meet the ozone standard.
Place a checkmark in the **CO** column to indicate actions that states must take in areas that do not meet the carbon monoxide standard.

O_3	CO	Actions Required to Meet NAAQ Standards
		"enhanced" vehicle I/M (inspection/maintenance) programs that more closely simulate driving conditions
		use of oxygenated fuels during the winter months
		transportation policies that offset any increase in vehicle miles traveled

29. List three types of clean-fuels. ______________________________
____

30. List advantages and disadvantages of clean fuels.
Advantages: ______________________________
____
Disadvantages: ______________________________
____
____

31. Some experts suggest that clean fuels will not solve the air pollution problem in urban areas. Identify other ways to reduce air pollution.
(1) ______________________________
(2) ______________________________
(3) ______________________________
(4) ______________________________

32. Identify the characteristic of mass transit systems that are the most successful in attracting passengers. ______________________________
____

33. Diesel-powered buses are contributing to air pollution. Trucks and buses are the major sources of ______________________________.

34. Identify two changes that can reduce pollutants from diesel-powered buses.
(1) ______________________________
(2) ______________________________

Name ________________________ Date ________ Class Period ____

Unit II — 6.3 Asbestos: Everybody's Problem

Introduction [191]

1. Identify three places where a person might be exposed to asbestos.
   (1) ______________ (2) ________________(3) __________________

What Is It? [191]

2. **Asbestos** refers to a group of **(synthetic fibers/natural minerals)**.

3. Identify the properties of asbestos that made it an important material used in construction. (1) ____________________ (2) ____________________
   (3) __________________________ (4) __________________________

4. Explain why asbestos might have been used in each of the following situations:
   A. It was sprayed on the walls in the music room to ____________________
   B. It was used in the blankets made for use on ships because it provided good ________________________ and it was not ______________________.
   C. It was used to wrap pipes because it provided ______________________.

How Can It Affect Your Health? [191]

5. Asbestos fibers that become lodged in the lungs make breathing difficult and cause a condition called ____________________.

6. Asbestos may cause ___________ cancer and __________________ cancer in the lining of the chest or abdomen. Because it causes cancer, asbestos is classified as a ________________.

7. Some people who worked with asbestos during World War II have recently filed law suits against their employers. Why did they wait so many years?
   ____

8. Asbestos is not dangerous unless it is in this condition _______________. If a product made of asbestos crumbles, the asbestos fibers can enter the _________. The product most likely to contain asbestos that will escape into the air is one that is used for **(insulation/flooring)**.

Schools and Other Public Buildings [192]

9. To insure the safety of students and workers, the **Asbestos Hazard Emergency Response Act** required schools to:
   (1) ____
   (2) ____
   (3) ____
   (4) ____
   (5) ____

10. **(True/False)** Schools that don't take steps to address potential asbestos problems can be fined by the EPA.

11. **(True/False)** Asbestos has been found in government and privately owned buildings as well as multiple and single family homes.

Name ______________________ Date __________ Class Period ____

12. **(True/False)** Many of the buildings that contain asbestos were built during the 1980s.

Is There Asbestos in Your Home? [192]

13. The vinyl floor tile is very worn. What should be done? ____________________

____

14. You would like to paint a new wall with textured paint. Do you need to be concerned that the paint you buy may contain asbestos? ______________

15. You purchase a home that was built in 1960. You would like to remove a section of a ceiling to put in a stairway. Should you check first to see if the ceiling contains asbestos? ____ If it does, how should you proceed with your planned remodeling?

____

16. You have recently purchased a home with a furnace that is insulated with asbestos. Should you consider replacing it? ______________________________

17. You notice that a small section of insulation material covering some pipes in your basement is loose and crumbly. You are not sure whether this material contains asbestos. Who might be able to tell you if the product contains asbestos?

____

____

What is the best way to handle the problem if the material is asbestos?

____

____

What Should You Do If You Find Asbestos in Your Home? [193]

18. **(True/False)** Material containing only one percent asbestos is subject to federal regulations.

19. **(True/False)** Most materials containing asbestos should be removed.

20. **(True/False)** If in good condition, materials containing asbestos are not considered dangerous.

Safety Guidelines for Working with Materials that Contain Asbestos [193]

21. **(True/False)** Any material that contains asbestos should be removed.

22. **(True/False)** Before removing asbestos materials the area must be sealed off from other areas of the building.

23. How should you dress when working with material that contains asbestos?

____

____

24. **(True/False)** If possible dispose of clothing that is worn during removal of materials containing asbestos.

Name ______________________________ Date ________ Class Period ____

25. Before cutting material that contains asbestos, it must be ___________.

26. When asbestos is removed it should be placed in ________________________

27. A **(vacuum cleaner/wet mop)** should be used to clean the area after asbestos has been removed.

Disposal of Asbestos Wastes [194]

28. A truck containing materials with asbestos arrives at the landfill you operate. What procedures should you follow? ________________________________
____
____

29. A trash hauler notices that several bags set out for the regular trash collection are filled with materials from a remodeling job. Some of the bags are split open and the hauler suspects that the materials contain asbestos. How should the hauler handle the situation?
    ____ Leave the bags with a note to the owner and take the remainder of the trash.
    ____ Pick up the bags along with the remainder of the trash from the area.
    ____ Leave the bags and notify the EPA.

30 The landfill has been filled and no more wastes can be accepted. What must be done so that the asbestos buried there will not become a pollutant?
____
____
____

Introduction [196]

1. The basic building block of all matter is the ______________.

2. The center of an atom is the ________________. It is surrounded by small particles called __________________.

3. The joining of two or more atoms to form a compound, such as water, involves interactions between **(electrons/nuclei)** of the hydrogen and oxygen atoms.

4. Radiation is released when the ________________ of an atom disintegrates.

Types of Radiation [196]

5. Place **P** beside each type of radiation that is in the form of a particle and an **W** beside each type of radiation that is a wave of energy.
   _____ Alpha radiation _____ Beta radiation _____ Gamma radiation

6. Identify the type of radiation that is stopped by each of the following materials.
   A. __________ a layer of lead several inches thick.
   B. __________ a sheet of paper.
   C. __________ a layer of air several inches thick.
   D. __________ a layer of water several feet deep.
   E. __________ a layer of aluminum at least 1 mm thick.
   F. __________ the outer layer of your skin.

7. Which type of radiation would probably not damage the human body unless swallowed or inhaled? _______________

8. Which type of radiation is the most penetrating? _______________

9. In the United States, the buildings where nuclear reactors are located are made of concrete that is several feet thick. The building is designed to protect living organisms from ________________ radiation.

10. **(Ionizing/Nonionizing)** radiation causes the most damage in cells by affecting the position of the __________________.

11. Identify each of the following types of radiation as ionizing **I** or nonionizing **N**.
    _____ Alpha _____ Beta _____ Cosmic _____ Gamma _____ Solar

12. As the wavelength of radiation increases the energy level **(increases/decreases)**.

13. Place the following types of radiation in order of decreasing energy by assigning the number 1 to the type of radiation with the greatest amount of energy.
    _____ cosmic _____ gamma _____ visible light _____ x-radiation

Radiation From Natural Sources [196]

14. You are told that Iodine–135 is a naturally radioactive element. The ____________ is unstable, and it spontaneously disintegrates or ________________.

15. When a radium atom "decays" is becomes a ________________ atom. The decay process produces particles and/or waves of energy called ________________.

16. Rocks that contain Uranium–238 and radium are ____________________.

17. The amount of naturally occurring radiation is called the ____________________ level of radiation.

18. As the elevation increases, the level of background radiation increases due to the presence of more ______________ radiation.

19. Who receives more cosmic radiation—a person living in Denver, Colorado, or a person living in Miami, Florida? __________________

20. If the time in flight is equal, who receives more radiation—an airplane pilot or an astronaut? ____________________

21. What two factors determine the level of radiation in your home?
(1) ____
(2) ____
____

22. Which type of power plant releases more radiation during normal operation—a coal-burning power plant or a nuclear power plant? _______________________

Radiation from Human-Made Sources [197]

23. Human-made sources of radiation are more dangerous than natural sources because the radiation is more ____________________.

24. List three ways in which the use of radiation has made your life safer or easier.
(1)____
(2)____
(3)____

Biological Effects of Radiation [198]

25. Naturally radioactive substances normally enter the body as a part of ___________ or _____________.

26. **(True/False)** Humans are not radioactive unless they are treated with radioactive drugs.

27. Even with all the data available, scientists still find it difficult to determine the effects of low levels of radiation on the human body. Give two reasons for this.
(1) ____
(2) ____

28. Radiation damages the body by changing chemicals which affect the cell's ________________ or it may interfere with important ________________.

29. Identify the term that describes each reference listed below.
    A. Take 2 aspirins with a glass of water. **(dose/dose rate)**
    B. Following the accident, some people living near the Chernobyl nuclear power plant were exposed to 25 millirems radiation per hour from contaminants in the air. **(dose/dose rate)**
    C. A patient is exposed to 20 mrem of radiation for each chest x-ray taken. **(dose/dose rate)**
    D. The average person is exposed to 100 mrem background radiation each year. **(dose/dose rate)**

30. In each of the following pairs, select the situation in which the most damage will occur.
    A. **100 millirems/1000 millirems**
    B. **500 millirems per year/500 millirems per week**
    C. **Alpha radiation/X-rays**
    D. **muscle cells/skin cells**
    E. **adult/child**
    F. **person who is healthy/person with AIDS**

31. The period of time between exposure to radiation and the appearance of cancer is the ________________. If a person is exposed to a large dose of radiation, the latent period is likely to be very **(long/short)**.

32. Radiation doses below 1,000 millirem increase the **(certainty/probability)** of damage. Compared to other health risks, the risk from exposure to low levels of radiation is small.

33. A radiation dose of 1,000–10,000 millirems is most likely to affect **(an adult/a child/an embryo)**.

34. Due to the latent period required, exposure of an embryo to 1,000 millirems of radiation would most likely cause **(birth defects/development of tumors during childhood)**.

35. An accident exposed workers to 100,000 millirems of radiation. Blood tests showed a decrease in the number of __________ blood cells. The workers have an increased probability of developing ____________.

36. Some of the men who fought fires caused by the accident at the Chernobyl Nuclear Power Plant suffered from "radiation sickness." Some of these men later died, and some were successfully treated by a doctor from the United States. Assuming that the men were exposed to 350,000 millirems of radiation, what type of treatment allowed them to survive? _______________

37. A patient receiving radiation therapy for cancer wears a hat to hide the fact that his hair has fallen out and is often nauseous after each treatment. Based on this information, the dose of radiation which the patient is receiving is greater than ___________ millirems but less than ___________ millirems.

38. A dose of one million millirems of radiation causes death due to damage to the _____________ system. A dose of ten million millirems of radiation causes death due to damage to the ________________ system.

39. Is a dose of 1,000 millirems dangerous for an adult? ________________ Explain.
____
____

40. Airline pilots who regularly fly high-altitude New York–Tokyo flights receive 900 millirem per year cosmic radiation. If your friend was an airline pilot assigned to that route, and she told you that she had just learned she was pregnant, would you suggest that she request a change in her flight assignment during her pregnancy? ______ Explain. ____
____
____

Vocabulary Review

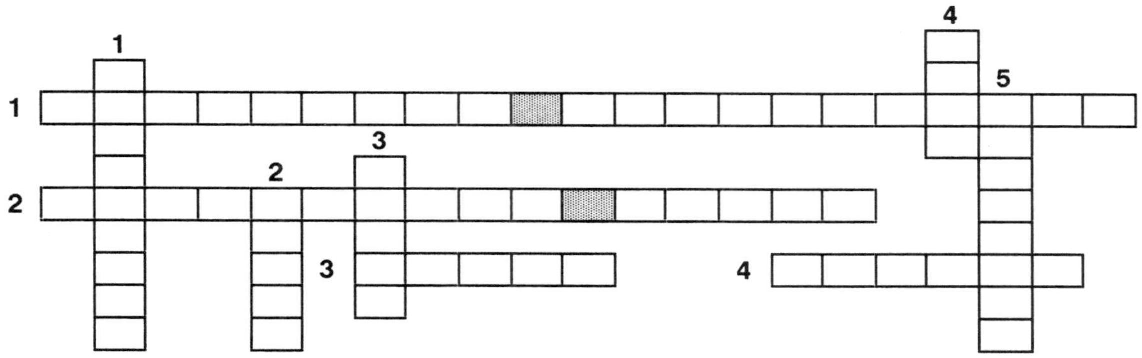

ACROSS:

1 An element with a nucleus so unstable that it disintegrates without any outside force.
2 The level of naturally occurring radiation.
3 Particles that can be stopped by the human skin, but if inhaled they are more damaging than other forms of radiation.
4 Waves of ionizing radiation that enter the earth's atmosphere from outer space.

DOWN:

1 Energy in the form of waves or particles that are released when the nucleus of an atom breaks down.
2 Waves of energy that have less energy than cosmic radiation but more energy than x-radiation.
3 A form of nonionizing radiation.
4 Type of radiation that can be stopped by a millimeter thick sheet of aluminum.
5 Includes all types of radiation that are capable of knocking electrons out of their normal position.

Danger at Home [202]

1. The Watras family moved out of their home because of:
   A. an accident at a nuclear power plant nearby.
   B. leaks in a natural gas pipeline.
   C. a naturally occurring radioactive gas.

2. **Radon** gas is a product of the "decay" of radioactive atoms of ______________.

3. Radium is one of a series of chemicals formed in the radioactive decay of ________________ found in rocks and soil.

4. Radon gas usually enters homes through ________________________________
   ____
   It may also be dissolved in ____________ pumped from wells and released into the air when you ____

5. The Watras house is built on top of _____________ rock that contains 50 ppm ________________ The _______________ house built to conserve ___________ and it trapped the ___________________.

Scientific Studies [202]

6. The granite beneath the Watras house was a part of a large granite rock formation called the _________________ that extends from southeastern _______________ into __________________ and ________________. Scientists studying the rock formation hoped to find valuable deposits of __________________.

7. The Department of Environmental Resources did not immediately investigate the high levels of radiation in the Reading Prong because of __________________
   ____

8. The purpose of the study done by the Pennsylvania Power and Light CO. was to see if air quality in homes was affected by ______________________________

   They found levels of radon gas that **(were/were not)** considered safe.

9. According to Dr. Sachs, the source of radiation that presents the most serious health risks is the radiation released from **(radon/the TMI accident)**.

The Invisible Threat [203]

10. Radon is a radioactive gas that "decays" into solid particles called _____________ ____________________.

11. The radioactive particles give off _____________ radiation which increases the risk of lung _______________.

12. Most studies of the health effects of radon gas have involved people who work in **(nuclear power plants/uranium mines)**.

13. Rank the following causes of lung cancer. ___ cigarette smoking ___ radon gas

14. What percent of non-smokers develop lung cancer? ___________ What percentage of lung cancer in non-smokers may be due to radon gas? __________

15. Uranium miners that smoke are **(twice/five times/ten times)** as likely to get lung cancer.

16. Radiation causes more damage in a smoker than a non-smoker. This effect of one pollutant intensifying the damage caused by another pollutant is called the

    ____

Are You at Risk? [204]

17. The unit used to measure the level of radon is ______________ per _________ of air. If a liter of air in your home has 2.2 atoms decaying per minute, the level of radon in your home is ______________________________.

    The EPA recommends that the level of radon in your home should be less than _______ picocuries per liter.

18. Radon gas is found in the soil, water, and air **(only in the East /throughout the U.S.)**. Have high concentrations of radon been found in the state where you live? ____________

19. The two factors that determine where homes have high levels of radon gas are:
    (1) ____
    (2) ____

20. According to the EPA, which homes should be tested for radon? _____________

21. The best place to put the detector would be in the _______________________ with doors and windows _______________.

    The test kit contains _____________ that adsorbs radon gas for a specific time or it contains a piece of ______________ that is sensitive to alpha particles.

Guarding Against Radon [205]

22. What measures were taken to radon-proof the Watras home?
    (1) ____
    (2) ____
    (3) ____

23. The EPA **(made radon reduction systems mandatory for new homes/proposed voluntary guidelines for builders)**.

24. For people living in the United States, radon trapped in homes is responsible for ________ percent of the average person's exposure to radiation.

    Do you think that building codes should require builders to install radon protection systems? ________________ Explain. _______________________________

    ____

    ____

Name ________________________________ Date __________ Class Period ____

Unit III — 1.1

Food . . . Before the Big Mac

The Hunters and Gatherers [209]

1. Plants provide humans three essential needs. These are: __________________, __________________, ____________________

2. Societies may be described by the way they get food. Food may be obtained by: ________________ and ____________________ or __________________ and ____________________ which is known as ______________________. If one hundred years represented the entire time that humans have been on this planet, the time that agriculture has been practiced would be represented by ________ year(s).

How Do We Know? [209]

3. An **artifact** is an object that has been **(changed by humans/preserved in its natural state)**.

   Classify the following items by placing an **A** in front of those that are artifacts.
   _____ arrowhead _____ mortar _____ smooth stone _____ woven basket

4. Which of the item(s) listed above could be dated using radiocarbon dating techniques? ____

5. If a clam shell found in a "dump" of an ancient Indian tribe has ½ of the carbon-14 found in the shell of a clam living today, how old is the clam shell found in the dump? ________________________

   If it has 1⁄16 of the carbon-14 found in the shell of a clam living today, how old is the clam shell found in the dump?________________________

The Beginning of Technology [210]

6. **Technology** includes ________________ and ________________ that make life easier. Give two examples of technology used in a hunting and gathering society. ____________________ and ____________________.

7. The volcanic rock used to make arrowheads that had edges like glass was **(obsidian/shale)**

8. The scientist who studies the technology used by early societies is a(n) ______________________. [Hint: see "How Do We Know?"]

Before Wheat and Corn [211]

9. In a hunting and gathering society, flour was sometimes made from these nuts which are produced by oak trees. ____________________

10. The bowl-shaped container in which they were ground was a ________________.

Name ______________________________ Date ________ Class Period ____

11. Identify the chemical that makes acorns bitter. ________________ The Indians poured water over the acorn flour to dissolve and remove the bitter flavor—a process called ________________.

12. An important technology in any society are containers that hold things. The containers in the hunting and gathering society were ________________. The design and weave depended upon the ________________.

Waterweed: Wild Rice [211]

13. Through the science of ______________ we have plants and fruits that are ___________. Harvest is easier because the fruits __________ at the same time, and the harvest is larger because plants have been bred to give a larger ________.

14. Identify the products of technology that are used in the harvest of wild rice.
____

15. The trait of wild rice that insures planting for next year's crop is ______________
____

16. Identify the process used for each of the following:
A. The hulls of wild rice are cracked by ________________
B. The hulls of wild rice are loosened by ________________
C. The hulls are then separated from the kernels by ________________

17. Compare the early technology with the more modern technology used in the processing of wild rice and other grains by completing the following chart.

Process	Early Technology	Modern Technology
parching (drying)		
threshing (loosen hulls)		
winnowing (cleaning, removing hulls)		

Corn: Tamed by Humans [212]

18. Identify three characteristics of corn that prevent the seeds from being scattered.
(1) ____
(2) ____
(3) ____

19. Why do seed catalogs suggest that corn be planted in rows arranged in a block instead of one long row? ____
____

What will happen to corn if humans no longer planted it? ________________
____

Name ______________________________ Date ________ Class Period ____

20. Archaeologist have found evidence that ____________ was the main cereal crop in early American civilizations.

21. List some examples of the evidence found by archaeologist that indicates the importance of corn to early American civilizations. ______________________
____
____

22. How did the cobs of corn found at the bottom of the trash heap in Bat Cave, New Mexico, differ from the cobs of corn at the top of heap? __________________
____

23. How does pod corn differ from pop corn? ____________________________
____

24. Scientists think that the growing of corn may have been the beginning of ________________ or agriculture.

Unit III — 2.1 Agriculture — The Development of Technology

Introduction [215]

1. ______________________ is the practice of growing crops for food. In early agricultural societies, meat was provided by ______________ and the gathering of ______________ and ________________ provided important vitamins and minerals.

Early Tools [215]

2. Which product of technology is <u>not</u> an important tool used in the practice of agriculture?
   A. arrowhead B. digging stick C. bones and shells D. machete

3. The ________________ was used to clear trees and brush. After they dried, the trees and brush were ________________.

4. Unlike the neatly tilled fields seen on farms today, the fields of the American Indians contained ________________ and logs.

5. Axes were made of _______________. Sticks were hardened by __________, and the blades of hoes were made of ___________, ___________ or _________.

Bigger Machines to Prepare the Ground [215]

6. When the hoe was attached to a beam and pulled by oxen, it was called a _______.

7. Write the name of each plow part on the line beside the correct letter.
   A. ______________ C. ______________ E. ______________
   B. ______________ D. ______________

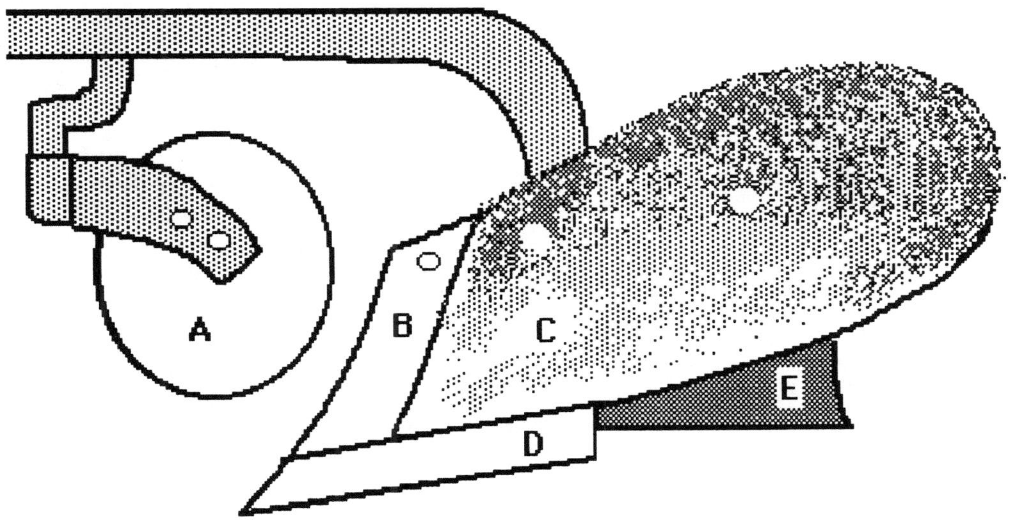

8. Match each of the following plow parts to their function.
coulter landslide moldboard plow share shin

___________ a large curved metal plate that turns the soil.
___________ attached to the bottom of the moldboard, a straight piece of metal that cuts through the soil.
___________ a straight piece of metal that serves as a guide for the plow.
___________ attached to the front edge of the moldboard, a replaceable piece of metal that cuts through the soil.
___________ located in front of the shin, a circular steel blade that cuts through the sod or debris on the surface.

9. Rearrange the following list so that the plows are in the order in which they were invented.
steel plow ____________________
cast-iron plow ____________________
wooden plow ____________________
gang plow ____________________

10. Identify the way that each of the following men improved the plow.

Charles Newbold ____
Thomas Jefferson ____
John Deere ____

11. The plow is both a good invention and a bad invention.
Describe the "good" changes that occurred because of the plow.

____

Describe the "bad" changes that occurred because of the plow.

____

12. Identify the two pieces of farm equipment that are used to prepare the seedbed after the field has been plowed. ____

13. Trace the development of the harrow by describing each of the changes that were made to improve the harrow.
The first harrows were simply made of ______________________________
These were replaced by harrows made of __________ with ________________
The prongs were replaced by ____________________
Modern harrows are made completely of ____________

Machines for Planting [216]

14. Seed was first scattered by ______________. Later the hand-cranked __________ ____________________ contained a fan that scattered the seed. The amount of seed scattered depends upon:
(1) ____ and
(2) ____

15. Today broadcast seeders are often used for planting **(large areas/small areas)**.

16. To plant large fields of grain, farmers often use a machine called a ________________________ The farmer puts seed and __________________ in long boxes. As the machine is pulled through the field, the seed and fertilizer fall to the ground through long ______________________________. A shallow trench is made in the soil by ___________ that are located at the bottom of the tubes. Drills designed for use on fields that have not been plowed or prepared for planting are called __________________ drills.

17. The first invention which made corn planting easier was the _________________. How is this mechanical device like the drill? ______________________________
____

Machines for Reaping [217]

18. Blades for cutting were first made of ________________.
These were later replaced with blades made of __________________.

19. How does a scythe differ from a sickle? ______________________________
____

20. A ___________________ was sometimes used with the scythe to ____________________________________. Grain was collected into bundles called ___________ and tied with __________ Bundles of grain were collected in groups called ______________.

21. What is another name for the reaper? ____________ The reaper __________ the grain and tie it into bundles with ________________. A reaper could do the work of ________ people.

22. What two events were important factors that helped increase the number of reapers sold?
(1) ____
(2) ____

23. Grain was placed on the wooden barn floor, called a ______________________, and beaten with a long wooden stick attached to a handle. The beater stick was called a ______________. A strong man could thresh seven or eight bushels of grain in ______ hours.

24. Eventually the flail was replaced by the _________________________.

25. The power for the first threshing machines was provided by ________________.
Later the ___________________ replaced the horses.

26. How many bushels of wheat could be threshed in one day using a flail? _________ using a threshing machine? ______________

27. The work done by the reaper and threshing machine is now done by the _________
____

28. The first corn pegs were made of __________________. Describe one of the corn pegs that was developed later. ____

29. The first machine that was used to harvest corn was the ____________________.

30. Today on small farms a ______________________ is often used to harvest corn while __________________________ are used on larger farms.

31. How do corn pickers remove the cob of corn from the stalk? ________________

____

32. What process is done by combine harvesters that is not done by the mechanical corn picker? ____

33. If you were assigned the task of shelling a bushel of corn, which of the four methods would you choose? ______________________________________

Do you think a shelling board would make the shelling easier? ______________

34. What mechanical device makes some models of hand-cranked corn shellers easier to turn? __________________________

35. How does the hand-cranked corn sheller remove the grains of corn from the cob?

____

36. What source of power has replaced the human hand for shelling corn today?

____

37. In what important way does a potato digger differ from a plow? ______________

____

More Food With Less Labor [220]

38. Why is more food available today even though fewer people are farming?________

____

39. Explain two ways in which the internal combustion engine made more food available for people.

(1)____

____

(2) ____

____

40. What percent of the total population lives and works on farms today? _________
How many people can one farmer feed? ___________

41. What advances in technology, other than machines, have made farmers able to produce more food?

(1) ____
(2) ____
(3) ____

42. What percent of jobs in the U.S. are provided by "off-the-farm" agricultural technology? __________

In Other Parts of the World [221]

43. What is the source of power for most of the farming that occurs outside the United States today? ____

44. Production of food in some countries is threatened by a lack of good __________.

44. Do you think that the best way to help people in other countries is by sending them some of our modern equipment? ________ Explain. ________________________

____

____

45. If you were given the money from a large benefit concert and were asked to spend it in a way that would most benefit the people in countries that do not have enough food, how would you spend it? ________________________________

____

Review

46. Identify the function(s) performed by each of the machines listed below:

______	flail	(1) cuts grain
______	reaper	(2) separates grain from other plant parts
______	binder	(3) both of these
______	threshing machine	
______	scythe	
______	combine harvester	

47. Identify the function performed by each of the machines listed below:

______	broadcast seeder	(1) preparing ground
______	reaper	(2) planting
______	combine harvester	(3) harvesting
______	threshing machine	
______	moldboard plow	
______	disk	
______	harrow	
______	grain drill	

Name ____ Date _________ Class Period ____

Unit III — 3.1 The Good Earth

Introduction [223]

1. **Technology**, the products and processes that improve life, allow farmers in the U.S. to produce more food than we need. This technology includes better ___________, human-made ____________________, and equipment that reduces the amount of __________________ needed to produce the food.

2. While most plants get nutrients from the ____________, some plants can be grown in a solution containing fertilizers. This method is called ____________________.

The Formation of Soil [223]

3. ____________, a thin layer of matter that covers the earth's surface, is created by the interaction of ____________, ____________, ____________, ______________, and ______________________________.

4. Soil formation is a **(fast/slow)** process. The depth of the soil covering the earth's surface **(is constant/varies)**. Soil is carried away by __________ and ____________.

5. Soil formation is faster in a **(cooler/warmer)** climate. The speed of chemical reactions increases as the ______________________ increases. The rate of soil formation is faster if the climate is **(dry/wet)**. ______________ is needed for chemical reactions.

6. The type of soil formed depends upon the __________________ and the type of ____________________ available. New York has **(less/more)** rain than Montana. Therefore, soil in New York has **(higher/lower)** concentrations of some minerals than soil made from the same type of rock in Montana. This is due to the process of _______________.

The Parent Material [224]

7. Match the words below with the statements that best describe them.
   (1) Minerals **(2) Rocks** **(3) Both**
   ______ A. are the parents of the soil.
   ______ B. are broken apart by physical and chemical forces.
   ______ C. become a part of the soil.
   ______ D. are chemicals that determine the color of the soil.
   ______ E. compounds that are found in rocks.

Weathering [224]

8. **Weathering** is the process in which _______________ are broken apart or ________________ are removed from them .

9. Chemical and physical forces that break rocks apart are called ________________ ________________.

Name ________________________________ Date __________ Class Period ____

10. Identify the **weathering agent** described in each statement.

_______________ A. breaks rocks when it forms in the cracks.
_______________ B. grows into cracks forcing rocks apart.
_______________ C. carried by moving water and glaciers.
_______________ D. acts as a sandblaster.
_______________ E. causes rocks to expand and contract.
_______________ F. produce chemicals which dissolve minerals from the rocks.

11. _______________ are produced by the actions of the weathering agents on rocks. These particles contain essential _______________ needed by plants.

12. Identify the approximate percent of the soil that is made of each of the following:

______ air ______ mineral particles
______ water ______ organic matter

Most of the soil consists of _______________.
The smallest portion of the soil consists of _______________.

Chemical Changes in Rocks [224]

13. Rocks containing FeO (ferrous iron) are _______________ in color. When exposed to the air, _______________ combines with the FeO to form Fe_2O_3 which is ___________ in color.

14. Match the following soils with the correct color.
(1) bluish-gray (2) red

_____ A. soil in a swamp
_____ B. soil that contains pores filled with air
_____ C. Soil that is well drained
_____ D. soil in a bog

15. Use the correct term to complete each sentence below:

acid	**carbon dioxide**	**limestone**
aluminum	**carbon monoxide**	**marble**
calcite	**clay**	**water**

The type of chemical that causes chemical weathering is an _____________. Carbonic acid is formed when _______________ (produced by the roots of plants) combines with _____________. Carbonic acid dissolves the mineral ___________ from _____________ or _____________ rocks. If carbonic acid dissolves too much _____________ from rocks, it is poisonous to the plant roots.

Caves are found in regions that have _____________ rocks. Evaporation of water leaves behind deposits of the mineral _______________. Deposits of minerals form _______________ that hang from the ceiling and _______________ that "grow" from the floor of the cave.

Texture of the Soil [225]

16. Mineral particles are classified by their _______________.

Name ______________________________ **Date** ________ **Class Period** ____

17. The property of soil that is determined by the size of the mineral particles it contains is called the ________________.

18. Which of the following particles is not usually considered to be a part of the soil?
A. clay **B. sand** **C. silt** **D. gravel**

19. Soil texture is determined by the percentage of ___________, ____________, and _____________ particles present in the soil sample.

20. Arrange the particles in question 19 in order from largest to smallest.
_____________, _____________, _____________

21. Which of the particles in question 19 cannot be seen without being magnified by an electron microscope? __________

22. Which of the particles in question 19 is the smallest particle that can be seen with the unaided eye? _________

23. The term used to describe soils that contain a mixture of sand, silt, and clay is ____________.

24. The soil is difficult to work if it has too many ___________ particles.

25. It is important to have clay particles in the soil because they increase the soil's ability to hold ____________ and ______________.

26. Nutrients enter the soil as a result of the processes of _________________ and _____________.

27. Nutrients are removed from the soil by plants and by the process of __________.

28. Clay particles have a _____________ charge. Because of this they attract and hold nutrients such as _____________, ________________, and ______________.

29. Soils with large amounts of clay are **(less/more)** likely to contain high levels of minerals than other soils.

Humus [226]

30. There is no soil on the moon because there is no __________ on the moon. Moon dust contains mineral particles, but it does not contain _____________.

31. **Humus** is partly ________________________________ that was once living or was _________________ by a living thing (waste).

32. Give two examples of materials that become humus.
(1) ____
(2) ____

33. How do earthworms, slugs, fungi and bacteria help create soil?_____________
____.

Name ___________________________________ Date _________ Class Period ____

34 Adding humus to the soil **(decreases/increases)** the soil's ability to hold **(nutrients/water/both)**. The type of mineral particle that holds water and nutrients is **(clay/sand/silt)**.

35. In order for humus to release its nutrients into the soil, the humus must _________.

Can There Be Too Much Humus? [226]

36. Rank the following soils beginning with the one that contains the most humus:

bog soil	(1) ______________
desert soil	(2) ______________
loam soil	(3) ______________
muck soil	(4) ______________

37. The color of muck soils is due to the high level of _________________ (organic matter) . Muck soils contain **(high/low)** levels of nitrogen and phosphorous. The productivity of muck soils is **(greater than/less than)** that of most other soils.

38. Muck soils are formed in areas that are _______________ and are usually too **(dry/wet)** to farm unless drained. When dry, muck soils may be lost due to the force of the _________________.

39. Since muck soils are not made from rock, they are often lack certain _________________.

Structure of the Soil [227]

40. Some ________________________ in the soil produce chemicals that glue mineral particles together. Other organisms produce chemicals, called ________________, that digest the organic matter or ________________.

41. Soil in which the organic matter and mineral particles are glued together to form clumps can be described as _________________ and ________________.

42. While the size of the mineral particles gives the soils its texture, gluing these particles together to form clumps gives the soil its _________________.

43. Two types of soils that lack structure are __________ and _________. Structure can be developed in these soils by adding _______________.

Spaces in the Soil [227]

44. It is important that the soil contain pore spaces because the soil must provide plant roots with ____________ and ____________. Normally________ to ________ percent of the volume of soil is pore space.

45. Describe how the plants will be affected in each of the following situations.
    a. Soil contains no pore spaces: ____
    b. Soil contains many large pore spaces: __________________________________
    ____

Name ______________________________ Date ________ Class Period ____

46. **Percolation** refers to the ability of water to ______________________
Identify the term that is used to describe the **percolation rate** of each of the following soils.

____________ A. 1 inch per hour
____________ B. 10 inches per hour
____________ C. 0.10 inch per hour
____________ D. 0.01 inches per hour

47. The size of the spaces in the soil depends upon the soil's ______________.
Soil with smaller particles contains **(larger/smaller)** spaces than soil with larger particles.

48. Identify the soil texture that has the best:
infiltration rate: **(clay/sand/silt)**
water-holding capacity: **(clay/sand/silt)**
aeration: **(clay/sand/silt)**

49. Soils that have good infiltration and aeration have **(good/poor)** water-holding capacity. Soils that have poor infiltration and aeration have **(good/poor)** water-holding capacity.

50. Identify the type of soil that dries out rapidly after a rain. __________ Identify the type of soil found in areas where water either runs off or stands on the surface but does not readily enter the soil. _________

51. When heavy rains cause a river to overflow its banks and flood a corn field, the plants die. Why?______________________________

52. Explain how adding humus will affect each of the following soils.
Clay soils: ______________________________
Sandy soils: ______________________________

Adding humus improves the soil's **(structure/texture)**.

A Soil Profile [229]

53. Match the terms with the descriptions given below.
(1) parent material (2) subsoil (3) topsoil

_____ A. layer that contains humus.
_____ B. layer that has the highest clay content
_____ C. layer that is the lightest color
_____ D. bedrock
_____ E. uppermost layer

Name ______________________________ Date __________ Class Period ____

VOCABULARY REVIEW

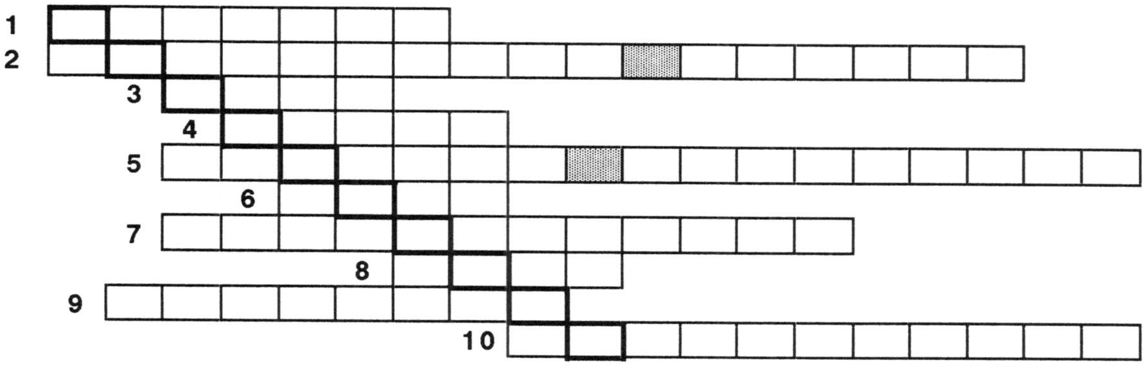

1 characteristic of soil determined by the size of the mineral particles.
2 chemical and physical forces that break rocks apart and/or remove minerals from them.
3 mineral particle that increases the soils ability to hold both nutrients and water.
4 partly decomposed organic matter
5 clay, silt, and sand
6 a mixture of mineral particles and humus that forms a thin covering over the surface of the earth.
7 the ability of water to soak into the soil
8 soil that contains less than 20 percent clay and nearly equal amounts of sand and silt.
9 process in which nutrients are dissolved and carried away by water
10 a method of growing plants in a nutrient solution

________________________ refers to products and process that improve our life.

The Land Provides [231]

1. Identify the technology used by the colonists that destroyed the land. ___________

2. The population density of the American Indians was **(greater/lower)** than the population density of people in Europe. The farming practices of the Indians were **(less/more)** destructive than the farming practices of the European settlers.

3. The plow turned the sod "upside down." The exposed soil was often carried away by __________ or __________. This process is known as **(erosion/leaching).**

The Debts Must Be Paid [231]

4. Who paid the transportation costs for many of the colonists who came to America? ________________________. The colonists who lived in New England repaid the company by providing ___________, ___________, and ____________ The colonists who lived in the South paid for their trip by growing ________________.

5. Describe the farming practice known as **monoculture.** ____________________
____

6. Explain how each of the following led to destruction of the land:
monoculture: ____
____

   cutting trees and plowing slopes: ____________________________________
____

Westward Ho! [232]

7. What were the colonists searching for when they moved to the Great Plains?
____

8. Buffalo and cattle are both grazers feeding on the lush grasses of the prairie. Explain why the prairie that had provided food for the buffalo for thousands of years was destroyed by the cattle. ______________________________
____

9. The native prairie grasses survive long periods without rain. Plants that can survive in this environment are said to be ______________________________

10. How do the native grasses differ from other grasses? ___________________
____

The Dirty Thirties [232]

11. Explain how each of the following events helped create the **"black blizzards."**
World War I: ____
Planting of corn and wheat: __________________________________
____
Weather conditions: ____

12. List the states which were a part of the **Dust Bowl** and locate these states on a map. ____

A National Plan to Save the Soil [233]

13. To help land owners protect the soil the **Soil Conservation Service** provided **(conservation specialists/financial assistance).**

14. The **Agricultural Stabilization and Conservation Service** was created to provide **(financial assistance/technical assistance).**

15. The **Soil Conservation Service** works to protect two natural resources: the ______________ and ______________.

Check the phone book under **"United States Government, Department of Agriculture."**
Is there a soil conservation district listed for your county? _________ Where are the offices located? ______________________________
What is the phone number? ________________
Who is responsible for planning and approving the soil conservation projects in your district? ______________________________
Are these people elected or appointed? ____________________

Types of Erosion [233]

16. Erosion is a **(human-made/natural process).**

17. Most erosion is caused by **(water/wind).**

18. Without human activities, erosion would **(occur at a slower rate/not occur).**

19. Match the descriptions given with the correct type of erosion.

(1) gully erosion (2) rill erosion (3) sheet erosion

_______ A. frequently seen on land with steep slopes and little or no plant growth.
_______ B. occurs on land with a very gentle slope and some plant growth.
_______ C. occurs on land with gentle slopes and little or no plant growth
_______ D. causes only small ditches.
_______ E. causes streams to be muddy even though a loss of topsoil is not obvious.
_______ F. the most obvious form of erosion.
_______ G. removes a thin layer of topsoil.

20. When traditional farming methods are used, the amount of topsoil lost from an acre of land in just one year is often **(less than 1 ton/more than 5 tons).**
If spread evenly over an acre of land, this amount of topsoil would form a layer **(1mm/1cm)** thick.
The thickness of this layer is best represented by a **(dime/nickel/quarter).**

21. Certain types of soils erode when they become saturated with water. The movement of a large mass of soil down a slope is called a landslide, but it is also known as _________ erosion or ____________________.

Slowing the Rate of Erosion [234]

22. The rate of replacement of topsoil is the ___________________________. During the 1930s the rate of erosion was **(greater/less)** than the tolerance value.

23 Match the conservation practices described in each of the following situations. Use each answer only once.

(1) conservation tillage
(2) contour farming
(3) crop rotation
(4) diversion terraces
(5) strip cropping
(6) waterway
(7) windbreak

___ A. strips of trees, shrubs, or tall grasses are planted in rows that are perpendicular to the direction of the prevailing wind.
___ B. ridges of soil are made along the contour of a slope.
___ C. a wide ditch is planted with a permanent grass cover.
___ D. after a crop is harvested, a different crop is planted in the field during the next growing season.
___ E. bands of close growing plants alternate with bands of crops planted in rows.
___ F. use of a chisel plow is:
___ G. horizontal plowing refers to:

24. Identify the conservation practice that would reduce erosion in each of the following situations. Use each answer only once.

(1) conservation tillage
(2) contour farming
(3) crop rotation
(4) diversion terraces
(5) strip cropping
(6) waterway
(7) windbreak

____ A. Large amounts of soil are lost when high winds blow across the large wheat fields on The Great Plains. During the winter the protective cover of snow is blown away and the young wheat plants are exposed to the cold. The soil conservationist recommends planting a:
____ B. A corn field is planted on a very gentle slope. This conservation method will help prevent rill erosion.
____ C. Gully erosion is occurring at the point where the water collects and runs off a field. A soil conservationist would recommend this conservation method.
____ D. Gullies are forming where traditional farming methods are being used in a field with steep slopes. To slow the speed of the flowing water, the soil conservationist would recommend construction of:
____ E. After the wheat is harvested, erosion by wind and water removes topsoil, plowing only increases the problem. The soil conservationist suggests this method of will reduce erosion and also reduce loss of moisture.

25. Indicate whether each of the following **increases (I)** or **decreases (D)** erosion.

_____ chisel plowing _____ moldboard plowing

26. Indicate whether each of the following is **increased (I)** or **decreased (D)** by no-till planting.

_____	A. compaction of the soil.	_____	F. water pollution.
_____	B. infiltration.	_____	G. evaporation of water.
_____	C. erosion.	_____	H. sheet erosion.
_____	D. energy required.	_____	I. wildlife.
_____	E. fertilizer absorbed by plants.	_____	J. pesticides used.

27. The method of farming that requires the greatest amount of planning is the **(conventional tillage/conservation tillage)**. Chemicals used with no-till planting are **(less likely/more likely)** to pollute the water supply.

28. Crops that are planted to hold the soil in place are called __________________. A grass which is often planted to hold the soil in place after the corn has been harvested is ______________. To hold the soil in place and add nitrogen to the soil, farmers might plant a _______________, such as clover, instead of a grass.

We've Learned Our Lesson — or Have We? [236]

29. In what ways was 1977 like the Dust Bowl years? ___________________________
____

30. Identify the two human activities that increased the rate of erosion.
(1) ____________________________ (2) ______________________________

31. Give three reasons why many farmers have abandoned many of the soil conservation practices that were developed after the Dust Bowl?
(1)____
(2)____
(3)____

32. **(True/False)** On some farms, more topsoil is being lost today than was lost during the Dust Bowl years. **(One-third/One-half/Two-thirds)** of the nation's farmland is losing soil faster than it is replaced.

33. The average farm will to lose one inch of topsoil in _________ years. If an average farm has eight inches of topsoil, and the rate of erosion is not reduced, how many years will it take to lose all of the topsoil? ______________________________

34. Explain why soil on the top of a hill is often lighter in color than soil farther down the slope. ____
____

35. Clothing and toys are sometimes imported from China. Explain how an Iowa farmer might be planting corn in soil that has been "imported" from China.
____
____

It's Not Just the Farmers [237]

36. List 5 human activities, other than farming, that increase the rate of erosion.

____

____

37. Which earth-moving machine has caused the greatest amount of erosion? **(bulldozer/moldboard plow)**
Which of the following yields more dollars per acre? **(cropland/grassland)**

38. Explain why big businesses and poor farmers are often "sodbusters" who do not practice soil conservation. ____

____

The Effects of Erosion [238]

39. Give three reasons why silt is considered a pollutant when it is carried in water.
(1)____
(2)____
(3)____

40. The productivity of the land has **(decreased/increased)** due to erosion. In spite of the population increase, the United States has been able to produce more food than the country needs. Give two reasons why.
(1) ______________________ (2) ______________________

41. Which of the following is a renewable resource? **(crude oil/soil)**

42. Explain how the rate of erosion is related to our dependence upon crude oil from the Middle East. ____

____

Irrigation Destroys the Soil [238]

43. Identify the technology that allows corn to be grown on The Great Plains.

____

44. Identify the source of water used to irrigate crops grown on The Great Plains.

____

45. The water used for irrigation is a **(nonrenewable/renewable)** resource.

46. Describe the relationship that exists between the Ogallala Aquifer and the fast food industry. ____

____

47. Give the cause of each of the following situations:
The Great Plains will become the Great American Desert.______________________

____

The Fertile Crescent is no longer fertile. ______________________

____

Much of California's irrigated farmland is no longer productive. ______________
____

48. What is the advantage of subsurface drip irrigation? ________________________
____

Why isn't subsurface drip irrigation used on more farms? ____________________
____

VOCABULARY REVIEW

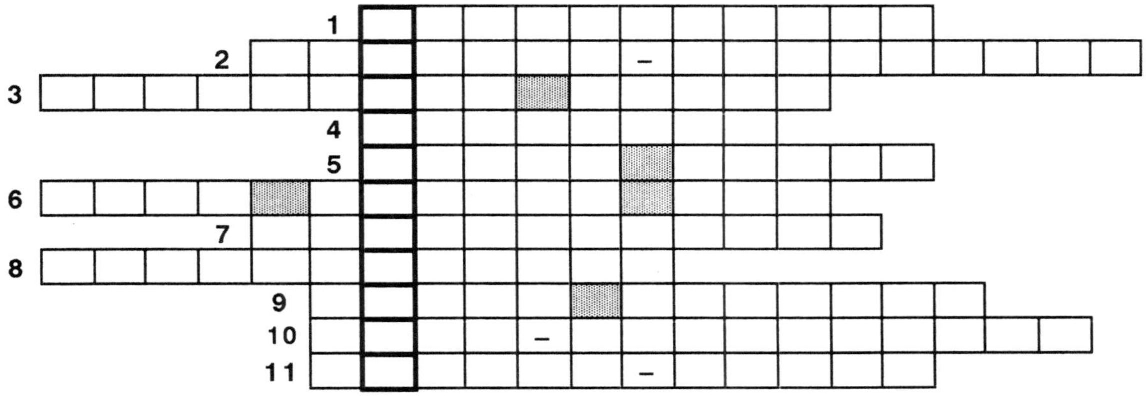

1. The practice of growing only one type of plant on a large area is:
2. Plants which can tolerate long periods without rain are:
3. The rate at which top soil can be replaced (formed) is the:
4. The aquifer which supplies water for irrigation on The Great Plains:
5. Legumes or grasses which are planted to hold the soil in place are :
6. A diagram which shows the location of different types of soil is a:
7. The build-up of mineral salts in the soil.
8. The amount of food produced per acre is the:
9. The type of erosion which occurs on steep slopes which have little or no plant growth.
10. A method of supplying water directly to the roots of plants through under-ground pipes.
11. A type of irrigation which uses large sprinklers that "walk" in a circle.

____________________ is the practice of growing only one type of plant on a large area.

Unit III — 4.2 A National Monument Created by Erosion

[page 243]

1. Refer to a map of the United States and locate South Dakota. Draw a sketch of the state and mark the location of each of the following features:
   A. Black Hills B. Badlands National Park C. Missouri River

2. On your reference map locate the following rivers:
   (1) Cheyenne River (2) White River (3) Bad River.

3. On the map drawn above, trace one of the possible paths a particle of clay might follow as it is carried away from the Badlands National Park during a heavy rain storm.

4. Number the events described below to show the order in which they took place.

____ The earth pushed upward draining the sea from the land.
____ South Dakota covered by the sea.
____ The Great Plains formed by layers of wind-blown mineral particles that eroded from the Black Hills and ash from volcanic eruptions that formed the Rocky Mountains.
____ Mud and organic matter deposited on the sea floor.
____ Movements within the earth pushed up a mass of granite rock covered with shale creating the Black Hills.
____ Material at the bottom of the sea changed into shale.
____ Weathering and erosion of the soft rock created the Badlands.

5. Identify the type of material that forms each of the following:

__________ rock covering the Black Hills that was formed by deposits in the sea
__________ rock in which the President's faces are carved at Mount Rushmore.
__________ material carried by water in the White River.

6. List four major weathering agents that are responsible for formation of the Badlands: (1) ____________________ (2) ____________________
   (3) ____________________ (4) ____________________

7. The type of erosion that forms deep grooves in the clay bluffs of the Badlands is **(gully/mass/rill/sheet)** erosion.

8. The rate of erosion in the Badlands is **(faster/slower/about the same as)** the rate of erosion in other parts of the United States.

9. The faces carved in Mount Rushmore preserve the memory of Presidents who made important contributions to the history of this country. What is the major purpose of establishing the Badlands National Monument? ____________________
____
____

Name ______________________________________ Date _________ Class Period ____

Unit III — 5.1 Can Technology Feed the World's Hungry?

The Population Connection [246]

1. Give the population reached in each of the following years.
   1800 ____ billion
   1930 ____ billion
   1960 ____ billion
   1974 ____ billion
   1987 ____ billion

2. The **doubling time** is the number of years required for a population to double its size. What was the doubling time required for each of the following populations?

   ______ **1 billion** [years required for the earth's population to reach 2 billion]
   ______ **2 billion** [years required for the earth's population to reach 4 billion]

3. In 1987 the earth's population was _____ billion. The current growth rate of the world's population is ________ percent.

   [To calculate the "doubling time," divide the number "70" by the percent growth rate.] At the current growth rate, what is the doubling time required for the 1987 population? ____________________

   The doubling time for the 1930 population was ________ years
   How does the predicted doubling time for the 1987 population compare to the doubling time required for the 1930 population?________________

4. The **natural increase** of a population is calculated by subtracting the ________________ from the births.

   In 1990 the birth rate in the United States was 16 per 1,000 and the death rate was 9 per 1,000.
   The natural increase was ________ per 1,000.

5. In 1990 the total United States population was 251 million (251,000,000). [This is 251,000 groups of 1,000 people.]

   Number of groups x Natural Increase per group = Natural Increase

   The **natural increase** in the U.S. population was ________________ people.

6. To determine the **growth rate** of a country, the effect of ____________________ must also be considered.

   ______________ + _________________ = _________________
   Natural Increase + Net Legal Immigration = Population Growth

Name ______________________________ Date ________ Class Period ____

7. To calculate the **percent growth rate** of the United States for 1990 use the following formula:

_____________ ÷ _______________ x 100 = _________
Population Growth ÷ Population at start of year = % Growth Rate

8. How does the growth rate of the United States compare to the growth rate of the world? ________________

9. Based on the information in the chart, identify the region of the world where some countries have reached a growth rate near zero. ________________

10. Based on the information in the chart, identify the region of the world that has a growth rate which is greater than 3 percent. ________________

11. If the world's population growth rate was 3 percent, how many years would it take for the 1987 population of 5 billion to double to 10 billion? _______ years.

How does this compare to the actual doubling time? ____________________
____

Why People Go to Bed Hungry [247]

12. In some countries wood is no longer a renewable resource because wood is the only source of ________________.

13. **(One-third/one-half/two-thirds)** of the earth's population depends upon wood as a source of fuel.

14. Identify two other natural resources that are affected by the increasing human population. ________________, ________________.

15. The population growth rate is **(decreasing/increasing)**, but the total world population is **(decreasing/increasing)** and the demand for food is **(decreasing/increasing)**.

16. Today in nearly **(one-third/one-half/two-thirds)** of world, there is frequently not enough food to feed the people.

17. What policy has been put into effect by the Chinese government in order to control the population growth rate? ______________________ What event during the 1960s led to this action? ________________

18. List four factors which result in a lack of food in some regions of the world.
(1) ________________________ (3) ________________________
(2) ________________________ (4) ________________________

19. The ability of the soil to absorb (infiltration) and hold water is reduced when the **(humus is/minerals are)** removed by erosion.

Name ______________________________________ Date _________ Class Period ____

20. Each year more land becomes desert. Number the events listed below to show the proper sequence of events.

_______ infiltration and water-holding capacity of the soil decreases.
_______ forests are cut to provide wood and/or land for farming.
_______ the land becomes a desert.
_______ erosion removes minerals and humus from the soil.
_______ population increases.

The World's Tillable Land [247]

21. The countries listed below produce most of the world's food. Circle those countries that do not produce enough food to meet their needs.

Canada China India United States
(countries of the former) Soviet Union

22. Identify the percentage of the earth's land that:

_______ cannot be farmed because of climate and/or slope
_______ is suitable for tilling
_______ can only be used for grazing or is covered by forests.

23. Compare the percentages of the earth's tillable land and the populations in the following countries:

	China	**North America**
Tillable land	percent	percent
Population	percent	percent

24. The amount of tillable land for each person on Earth in 1950 was ____________
The amount of tillable land for each person on Earth in 1992 was ____________
In the future the amount of crop land per person will **(decrease/increase/remain the same)**.

25. List factors that determine the productivity of an acre of tillable land.
(1) ______________________________ (2) ______________________________
(3) ______________________________

26. As a part of a new government program you are allowed to select an acre of land. You will be expected to grow all of your own food on this acre.
(1) Give the location of the acre you would select. __________________________

(2) What foods that you now eat would no longer be available? _______________
____

Japan [248]

27. Japanese farms are **(larger/smaller)** than farms in the United States.

28. Japanese farmers have **(higher/lower)** yields per acre than U.S. farmers.

Name ______________________ Date __________ Class Period ____

29. Place a checkmark beside each technology listed below that is used on Japanese farms:
    ____ special varieties of seeds
    ____ large quantities of fertilizer
    ____ large mechanical equipment

30. Which of the technologies listed above will cause a decrease in productivity?
    ____

The Green Revolution [248]

31. Place a checkmark beside those statements that are true.

    Solving the problem of world hunger is dependent upon:
    ____ the developing countries being able to grow more of their own food.
    ____ developing countries importing farm equipment from other countries.
    ____ improving the methods of farming in developing countries.
    ____ farmers from developing countries coming to the United States to see how food is grown here.
    ____ scientists going to developing countries to help farmers.
    ____ developing countries controlling the population growth rate.
    ____ the development of new varieties of seeds

32. What technological development resulted in the "Green Revolution"?
    ____
    ____

India [249]

33. Why didn't the Green Revolution help the poor Indian farmers?
    ____
    ____

34. Many people in India cannot afford to buy coal or oil for cooking. Identify the two major sources of fuel and explain how their use decreases the productivity of the land.

    Fuel: __________ Productivity decreased because the __________
    ____

    Fuel: __________ Productivity decreased because the __________
    ____

Bangladesh [249]

35. Describe the low-cost irrigation technology that is increasing the production of food in Bangladesh. ______________________________
    ____
    What is the source of energy for this new technology? __________

36. How does the population growth rate of Bangladesh compare to the population growth rate of most other countries? ______________________________

Name ______________________________ Date ________ Class Period ____

More Protein Needed [250]

37. Unlike many Americans, **(one-fourth/one-half/two-thirds)** of the world population does not get enough protein.

38. Describe the effects of a lack of protein in the diet of
Children: ______________________________
____
Adults: ______________________________
____

39. Identify the food that supplies much of the protein for the people in Tanzania.
____

40. To increase the protein in their diet, Tanzanian farmers are being taught to raise a type of fish called ____________. The fish are raised in ______________.
They eat ______________________________
Their fast rate of growth allows them to be eaten within _____ months.

41. Explain how the Green Revolution decreased the amount of protein in the diet of people who grew rice. ______________________________
____

42. **(Animals/Plants)** are the main source of protein in the diet of most people. Which plant is a better source of protein? **(soybeans/wheat)**

The United States — Always A Land of Plenty? [250]

43. The Green Revolution produced improved varieties of ____________ which is a major crop grown in the U.S.

44. List two ways that soybeans are used by people in other countries.
(1) ______________________________
(2) ______________________________

45. Place a checkmark beside the <u>one statement in each pair</u> that is true:

___ The ability of the U.S. to produce enough food is unlimited because of the large amount of farmland available in the U.S.

___ The ability of the U.S. to produce enough food to meet the future demand depends upon development of varieties that produce higher yields.

___ Some of the land in the U.S. that is now covered by forest could be converted to cropland.

___ All of the land in the U.S. that is classified as tillable is currently producing crops.

___ The farmland conservation act of 1970 prohibits prime farmland from being developed or used for any purpose other than farming.

___ Some of the nation's best farmland is currently being developed.

Name ______________________ Date __________ Class Period ____

Preserving Farmland [251]

46. How can tax rates encourage landowners to maintain land as farmland?

____

____

47. How can farmland be protected from development while at the same time allowing the farmer to obtain the highest price?

____

____

____

48. How would the value of the development rights be determined?

____

____

____

The Hamburger-Rain Forest Connection [251]

49. List three reasons why countries, like the U.S. and Canada, which grow a surplus of food still import large quantities of food?
(1) ____
(2) ____
(3) ____

50. Rain forests in Central and South America are being cut and burned in order to make room for ______________________ .

51. Due to the high temperatures and large amount of rainfall, the organic matter in the tropical rain forest decays (**rapidly/slowly**).

The roots of (**grasses/tropical trees**) are adapted to quickly absorb the nutrients released by the decaying matter.

Once the rain forest is cut, ____________ are planted. Since the roots of the ____________ do not absorb the nutrients quickly, they are dissolved in water and ____________ from the soil.

52. What is the carrying capacity of 10 acres of newly cleared rain forest? [How many cows will it support?] _____ What is the carrying capacity of this land five years later? ______________________

53. Identify four limiting factors that cause the growth of grass to decline after the first year of growth.
(1) ______________________ (3) ______________________
(2) ______________________ (4) ______________________

54. What is the largest single cause of the decline in tropical rain forest?

____

55. List three foods that your family buys which might be made from beef grown in Central America.

____

Name ______________________ Date __________ Class Period ____

56. How does the amount of beef eaten by Central Americans compare to the amount of beef eaten by North Americans? ______________________

____

57. Do you think it would be possible to save the rain forest habitat by selling "development rights"? ______________ Explain.

____

____

____

If the Central American farmers sold the development rights to the rain forest that remains, how would this affect you? ______________________

____

How would it affect the diet of Central Americans? ______________________

____

The Sea as a Source of Food [252]

58. Nearly **(one-third/one-half/three-fourths)** of the Earth is ocean.

59. Technology has increased **(the ability of the ocean to produce fish/the ability of humans to catch fish)**.

60. The capacity of the oceans to produce fish is **(limited/unlimited)**. The greatest threat to edible species of fish may be ______________________.

60. How are land developers threatening the fish populations? ______________________

____

61. What evidence of the **edge effect** can be seen in the ocean? ______________________

____

62. Even though pollution along the shore may not kill the organisms we use for food, we might not be able to use them as food. Explain:

____

____

Can the Earth Feed 11 Billion People? [253]

63. If a population of 11 billion people does not exceed the earth's carrying capacity, why are some people dying of hunger?

____

____

64. What steps should we take to insure the food supply for the future?
(1) ______________________
(2) ______________________
(3) ______________________

Name ______________________________ Date __________ Class Period ____

VOCABULARY REVIEW

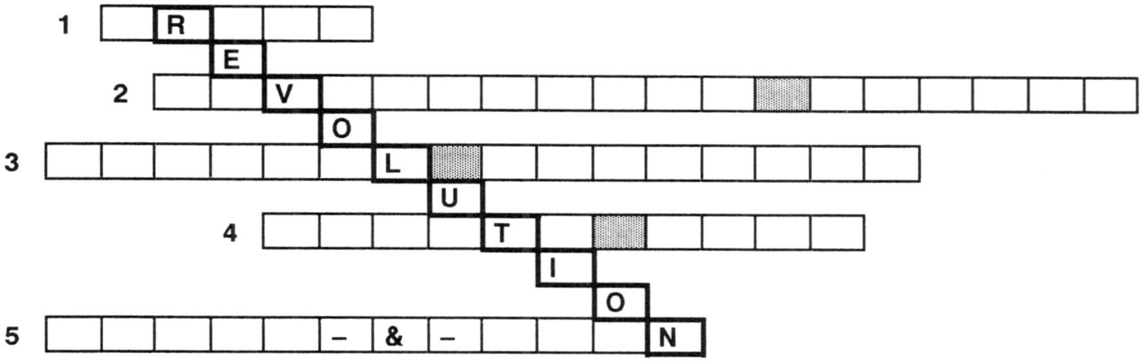

1. The revolution that produced new higher-yielding varieties of wheat and rice.
2. A restriction is placed on the deed to insure that the land is always be used for farming if the farmer sold the:
3. is due to more births than deaths.
4. the percentage increase in the Earth's population for a given year
5. type of agriculture practiced in regions where tropical forests grow

Unit III — 6.1 Fertilizers: Organic versus Inorganic

Introduction [256]

1. Soils in which plants grow well are said to be _____________.

2. Identify two types of natural materials that early farmers added to the soil to make the plants grow better. (1) _______________ (2) __________________

3. To prove that plants need certain chemicals to grow, scientists grew the plants in ______________ or ______________.

4. Soils that are "worn out" are lacking chemicals called __________________.

5. Nutrients are recycled by the process of ___________, but some nutrients are removed from the ecosystem by the processes of ______________ and ______________.

6. List two ways in which farming "robs" the soil of its nutrients.
   (1) ____
   (2) ____

Essential Nutrients [256]

7. Scientists have determined that plants need to obtain at least ______ chemicals from the soil.

8. Identify three other chemicals that plants need for photosynthesis and respiration.
   (1) ________________ (2) _________________ (3) __________________

9. Identify the three primary nutrients. (1) ________________ (2) ______________
   (3) ___________________

10. Identify the three secondary nutrients. (1) ______________ (2) _______________
    (3) ___________________

11. Humans need only very tiny amounts of some vitamins and minerals. Plants need only very tiny amounts of some elements. These elements are called ___________________.

12. Place **P** beside those nutrients that are primary nutrients.
    Place **S** beside those nutrients that are secondary nutrients.
    Place **M** beside those nutrients that are micronutrients.

_____ calcium	_____ boron	_____ phosphorous	_____ nitrogen
_____ iron	_____ magnesium	_____ manganese	_____ copper
_____ zinc	_____ potassium	_____ sulfur	_____ chlorine

The Primary Nutrients [257]

13. The primary nutrient that is most often the limiting factor for plant growth is _________________.

14. Plants that are lacking nitrogen will grow (**faster/slower**) and have (**dark green/yellow green**) leaves.

15. Place an checkmark beside the two forms of nitrogen that most plants can use.
_____ **NH_4^+(ammonium ion)** _____ N_2 _____ **NO_3^- (nitrate ion)**

16. Some plants remove large amounts of nitrogen from the soil. Identify two other processes that are responsible for the loss of nitrogen from the soil. [remember the odor from a bottle of ammonia] (1) _______________ (2) _________________

17. Farmers must be aware that cattle can be poisoned if the weather is (**dry/wet**) and too much (**nitrogen/phosphorus**) is present.

18. Phosphorus and potassium are needed by plants for proper _______________ and resistance to ________________.

19. Identify the primary nutrient that is the cause of each of the following conditions. Use the proper symbol: **Nitrogen (N)** **Phosphorus (P)** **Potassium (K)**

_______ if too much of this nutrient is present, plants have tall stems that break easily.
_______ if this nutrient is lacking, plants produce fewer flowers.
_______ if too much of this nutrient is present, plants are more easily damaged by frost.
_______ fruit does not ripen properly without this nutrient.
_______ this nutrient is needed for good root growth.
_______ too much of this nutrient results in large plants, but they do not have a good flavor.
_______ purple edges on the leaves of a corn plant tell a farmer the plant did not get enough of this nutrient.
_______ rows of kernels on corn cobs are crooked and sometimes incomplete if this nutrient is lacking.
_______ kernels do not fill out resulting in a small ear of corn that is sometimes called a "nubbin".

20. Purple-tinged leaves are not always a sign of a nutrient deficiency. Some plants are normally purple while others develop a purple-tinged leaf when the weather is _________________.

21. Firing or browning of the edges of plant leaves may result from a lack of the nutrient _______________ or, in some plants, it indicates the lack of _________________.

22. The leaves are a deep green color, but the plants are shorter than normal. This may indicate a lack of _________________ or ______________________.

23. Unless the soil has the proper __________, nutrients may not be in a form that the plants can use.

Soil Testing [258]

24. Give two reasons why farmers, gardeners and orchard owners are concerned about the level of nutrients in the soil.
(1) ______________________________
(2) ______________________________

25. Give the advantage(s) of each of the following soil tests:
State University laboratory: ______________________________
____
Private laboratory: ______________________________
Soil Test Kit: ______________________________

Choice of Fertilizers [258]

26. A substance added to the soil to supply one or more plant nutrients is a ______________ If it is changed chemically it is called a ______________ fertilizer.

27. Fertilizers made from dead organisms or their waste products are called ______________.

28. Bone meal and rock phosphate are both (**organic/synthetic**) fertilizers. Superphosphate is a synthetic fertilizer because it has been changed (**chemically/physically**).

29. Since the 1930s synthetic fertilizers have been made from ______________.

30. Straw, used for animal bedding, absorbs ___________ which is a good source of nitrogen. It is a (**natural/synthetic**) fertilizer.

31. Urea, which is made using atmospheric nitrogen and natural gas, is a ______________ fertilizer.

32. Identify three technological advances that increased the yield and quality of crops.
(1) ______________ (2) ______________
(3) ______________

An Unlimited Supply? [259]

33. The supply of synthetic nitrogen fertilizer is (**limited/unlimited**) because these fertilizers are made from a (**nonrenewable/renewable**) resource.

34. Synthetic nitrogen fertilizer is dependent upon the supply of ______________
____

35. Explain how war might affect the availability and cost of synthetic fertilizers.
____
____

36. ___________ are the source of all primary and secondary nutrients except nitrogen.

37. Few rocks contain the primary nutrient (**phosphorus/potassium**). Most of the rocks that contain this nutrient are located in two countries. Circle the two countries where these deposits are found.

Africa **Canada** **Russia** **South America** **United States**

38. Crops grown for food require (**less/more**) phosphorus than the grass or trees that naturally grew in soil where we now grow food crops.

39. Describe two pathways that phosphorus might travel on its way from the rock to the ocean.
(1) ____
____
(2) ____
____

40. Does potassium follow the same pathway as phosphorus? ____________
If so, why aren't scientists as concerned about the loss of potassium as they are about the loss of phosphorus? ______________________________
____

The Organic Way [260]

41. In what way is the organic garden like a forest ecosystem? ________________
____

42. Bacteria and fungi need the nutrient __________________ during the process of decay. After harvest when plant material is tilled into the soil, the amount of this nutrient will (**decrease/increase**).

43. To speed the process of decay, gardeners make ______________. A pile is made by alternating layers of ________ that contains bacteria, ___________ that contains nitrogen, and the ___________________________ that will be digested. The gardener must add __________ and turn the mixture to increase the supply of _________ needed by the bacteria.

44. Fungi help break down organic matter in the soil. They also help plants ________ ____________. When farmers use only synthetic fertilizer, the amount of organic matter in the soil (**decreases/increases**) and the fungi (**disappear/multiply**). As a result, (**larger/smaller**) amounts of fertilizer are needed.

45. Organic matter helps maintain good soil (**structure/texture**). Humus increases the ability of the soil to hold ______________ and ________________.

46. Give two examples of organic fertilizers produced from slaughterhouse wastes.
____

47. The solid material that remains after sewage is treated is called ____________. The sludge is mostly organic matter that conditions the soil or improves its ______________. It is also a source of ________________.

48. Sludge applied to farmland must be tested for ______________________ and treated to kill harmful ________________. Crops grown on this land should not be

consumed by ____________. To prevent water pollution, the amount of sludge applied to the land is determined by the amount of ____________ the crop will use.

49. If a farmer or gardener wants to increase the amount of organic matter in the soil, they might plant ____________ or _______. These crops are called __________ ____________ crops. In addition to adding organic matter, these crops decrease ______________.

Which is Better — Organic or Synthetic? [261]

Place an (**O**) in the blank if the statement applies to organic fertilizers, a (**S**) in the blank if the statement applies to synthetic fertilizers, and a (**SO**) in the blank if the statement applies to both.

_____ A. provide nutrients for plant growth.
_____ B. contain more concentrated chemicals.
_____ C. makes nutrients available to plants for a longer period of time.
_____ D. nutrient levels are standard and can easily be determined.
_____ E. is more expensive if purchased.
_____ F. contains low levels of nutrients.
_____ G. nutrients are not immediately available to the plant.
_____ H. people often apply too much.
_____ I. adds nutrients and increases the humus content of the soil.
_____ J. nutrients are more easily leached from the soil.

VOCABULARY REVIEW

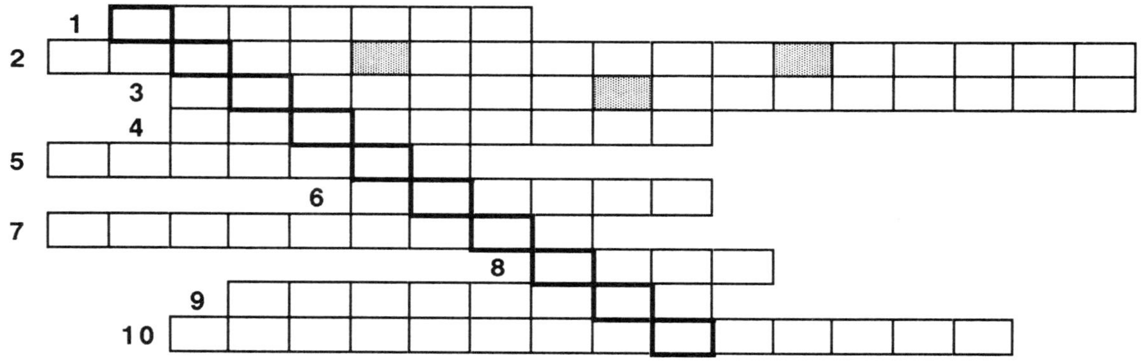

1. Soils where plants grow well are said to be:
2. Crops grown for the purpose of adding organic matter to the soil are:
3. Plants require large amounts of nitrogen. Nitrogen is a:
4. The nutrient needed for proper development and ripening of fruit is:
5. A term used to describe natural fertilizers even though they may not contain carbon.
6. Organic waste from sewage treatment plants is called:
7. Fertilizers made by chemically changing the nutrients so that they are readily available for plants.
8. Plants need very tiny amounts of this element. If present in large amounts it may prevent the growth of seedlings.
9. The nutrient which is most often the limiting factor for plant growth.
10. Those nutrients needed by plants in very small amounts are called:

Name ______________________________ Date ________ Class Period ____

Unit III — 7.1 Pests and Pesticides — They Both Cause Problems

Introduction [264]

1. The word "pest" comes from the Latin word for ______________.

2. Give the reason(s) why each of the following organisms is classified as a pest.
   Rats: ______________________________
   Fleas: ______________ ______________
   Mosquitoes: ______________________________
   Certain fungi: ______________________________
   Grasshoppers: ______________________________
   Boll Weevil: ______________________________

3. More than one fourth of the people living in Europe in the 14th century died of the disease known as the ______________. It is caused by ___________ that are carried from rats to humans by __________.

4. The two diseases that have killed more people than all the wars described in history books are: (1) ______________ (2) ______________

5. Identify the pest that:
   ______________ destroyed the potato crop in Ireland in the 1840s.
   ______________ was probably the worst insect pest the early settlers faced in the western U.S.
   ______________ entered the U.S. from Mexico in the 1890s.
   ______________ was the major insect responsible for the destruction of rangeland and grassland in the Western U.S. in 1985.
   ______________ was the major insect pest that destroyed the cotton crop in Coffee County, Alabama, in the early 1900s.

6. Place a checkmark beside the statement that is true.
   ___ The boll weevil and the boll worm are the only two major pests of the cotton crop.
   ___ There are more than twenty five pests that attack the cotton crop.

Miracle Chemicals [265]

7. In each of the following terms, "-cide" means to ____________.
   Give the target organism for each of the following chemicals.
   Avicides: __________ Insecticide: __________ Fungicide: __________
   Herbicide: ____________ Nematicide: ____________
   The term that includes all of the chemicals above is ____________

8. The first pesticides were ______________________.
   The Bordeaux mixture and other compounds containing the mineral __________ are still sometimes used to kill ____________.
   Wine makers use compounds containing ____________ to kill bacteria and wild yeast on fruit.

9. Identify the natural pesticides that are made from:
   chrysanthemums __________________ tobacco plants __________________

legumes growing in East Indies ____________________

10. Identify the (2) major disadvantages of the natural pesticides.
(1) ________________________ (2) ____________________________

11. Insects were used to test chemicals that were being developed for chemical ____________________
These chemicals affected the body's ____________________ system.
The two types of chemicals being tested were:
____________________ and ____________________.

12. Place a checkmark beside the phrase in each pair below that best describes pesticides that are **chlorinated hydrocarbons**.

cheap	expensive
killed many kinds of insects	killed only a few types of insects
easy to use	use required extensive training
easily washed away by rain	weren't washed away by rain
long lasting	are quickly broken down
not toxic to humans	toxic to humans

13. The first chlorinated hydrocarbon used as a pesticide was:
(chloroform/dichloro-diphenyl-trichloro-ethane/methane)
It is more commonly known as ____________.

14. Chlorinated hydrocarbons contain the elements ______________, ______________ and ______________. Scientists make different hydrocarbons by changing the __________________ and __________________ of the atoms.

15. More than ____________ different pesticides are produced by chemical companies. Most of these are used by _______________, but _______ percent of homeowners use pesticides for their homes and ____________. Most of the pesticides used today are **(herbicides/insecticides)**.

16. Match the insecticides listed with the proper classification.

____ A. DDT
____ B. Parathion
____ C. Pyrethrum
____ D. Chlordane
____ E. Nicotine Sulfate
____ F. Heptachlor
____ G. Sevin®

(1) carbamate
(2) natural organic
(3) organochloride
(4) organophosphate

Not All Are Pests [266]

17. About _________ of the 82,500 insects estimated to occur in North America are considered to be pests. This means that only _____ of every _____ insects is a pest.

18. Identify three ways in which some fungi are helpful to humans.
(1) ____
(2) ____
(3) ____

Name ______________________________ Date ________ Class Period ____

19. ______________ of our diet consists of crops that are ________________ by honey bees. The major food in the diet of dairy cattle is ____________. Explain how a decrease in honeybees might cause a decrease in milk production.
____
____
____

20. Before adding chemicals to the environment, it is important to study their effects upon organisms in the _________ as well as _________, _________, _________ and _________.

Monoculture — An Invitation to a Picnic [267]

21. List three places where monocultures can be found.
_____________, _____________, _____________

22. How does the practice of monoculture increase the population of insects?
____

Problems Caused by Pesticides [267]

23. When synthetic pesticides first appeared:
Doctors imagined a world with: ______________________________
Farmers imagined a world with: ______________________________

24. List two significant benefits of the use of synthetic pesticides.
(1) ____
(2) ____

25. Without the use of pesticides, at least _________ percent of a farmer's crop is lost before harvest. Another _________ percent is lost while the food is in storage. This means that only _________ percent of the food produced is available for human consumption.

26. Why did Rachel Carson write the book *Silent Spring* ? ________________
____

SUPERBUGS [268]

27. An insect that survives when sprayed with an insecticide is said to be _________. This ability to survive the spray is due to changes in the insect's ____________
____

28. Genetic changes make the insect immune or ____________ to the pesticide. When insects that are resistant mate with other resistant insects, the offspring are ____________ .

29. Increasing the number of times a pesticide is sprayed during a season will **(decrease/increase)** the number of resistant insects.

30. Most pests will become resistant to a pesticide within _____ years, but some may become resistant after only _____ season.

Name ______________________________ Date ________ Class Period ____

31. Identify the number of insect species that:
• were resistant to pesticides in 1965. _____
• are resistant to pesticides today: _____
This number is expected to **(decrease/increase)**.

32. Explain why insects develop resistance to insecticides more quickly than weeds develop resistance to herbicides. ______________________________
____

PERSISTENT PESTICIDES [268]

33. Pesticides that break down easily are said to be **(biodegradable/persistent)** pesticides.

34. Place a **P** beside those items that are persistent and a **B** beside those items that are biodegradable.

_____ most plastic	_____ organophosphates	_____ DDT
_____ paper	_____ chlorinated hydrocarbons	_____ heptachlor

35. The pesticide most frequently found in fruits and vegetable grown in California during a 1984 survey was ________. It had been _____ years since this pesticide was banned in the U.S.

36. Even though DDT has never been used in Antarctica, it is found in organisms living there. Explain how the organisms became contaminated with DDT. ____________
____

It is likely that biodegradable insecticides will be found in Antarctica? ________
Would you expect to find biodegradable pesticides in your body fat? ________

37. Give two reasons why farmers and scientists thought that persistent pesticides were better than biodegradable pesticides.
(1) ______________________________
(2) ______________________________

BIOACCUMULATION [269]

38. Chlorinated hydrocarbons are easily dissolved in **(fat/water)**.

39. Why did farmers and scientists think that this was an advantage?
____

40. What happens to DDT and other chlorinated hydrocarbons when they are carried into a lake with runoff during a storm? ______________________________
____

41. Because cell membranes are made of protein and ________, chlorinated hydrocarbons are easily absorbed through the cell membranes of aquatic organisms.

42. Most organisms do not have the _____________ needed to break down pesticides. Pesticides that are not soluble in water are stored in the body ________. If they are easily dissolved in water they are excreted through the ________________.

Name ______________________________ **Date** ________ **Class Period** ____

43. More and more pesticide accumulates in organisms at the bottom of the lake. As these organisms are eaten by other organisms the pesticide becomes concentrated in the predators. This is called ____________________.

44. List two ways that DDT may affect the reproduction of some fish-eating birds.
(1) ______________________________
(2) ______________________________

45. Before DDT was banned, populations of fish-eating birds, such as eagles and ospreys, were **(decreasing/increasing)**. Since the ban on DDT the populations are ______________.

46. A song bird that was affected by DDT was the ____________. The DDT caused the **(death of adult birds/eggs to break)**.

47. Draw a food chain that includes the robin and shows the pathway of the DDT.

48. Explain why people, robins and fish-eating birds have more pesticide stored in their body than other organisms such as crabs and earthworms? ______________
____
____

PESTICIDES CREATE PESTS [271]

49. If pesticides are compared to guns, ________________ pesticides kill in the same way that a rifle kills while ________________ pesticides kill in the same way that a shot gun kills.

50. Select the type of pesticide, **selective** or **broad-spectrum**, that most closely matches the description given.
______________ kills only one type of organism.
______________ kills many beneficial insects.
______________ its use sometimes makes a minor pest become a major one.
______________ actually increases the pest population by killing the predators.
______________ upsets the natural rise and fall of insect populations.

51. Refer to the * beside each number on the graph in the graph to answer the following questions.
*1—What is happening to the pest population? ________________
*2—Is the predator population changing? ________________
If not, give a reason why there is no change.
*3—What is happening to both the pest and predator populations? __________
What caused this change? ______________________________
*4—Which population is increasing? ____________

52. The use of a broad-spectrum pesticide leads to an increase in the pest population because the pesticide ______________________________.

Name ______________________________ Date __________ Class Period ____

Unit III — 7.2 A Lady Named Carson

[page 274]

1. The name of the book written by Rachel Carson is ______________________ It was published in 19_____.

2. Miss Carson had been trained as **(an author/a scientist)**.

3. What issue was discussed in the book? ______________________________
____

4. Miss Carson wanted to get information to two groups of people. These two groups were ______________________ and ______________________.

5. An investigation by an independent scientist showed that Miss Carson's statements **(were/were not)** supported by facts.

6. How did Miss Carson's book differ from other books written about pesticides?
____
____

7. In addition to chemical companies, the unrestricted use of pesticides was promoted by the United States ____.

8. Place a checkmark beside the statement in each pair of statements that best describes the position that Miss Carson supported.

___ Citizens should be told when the area where they live and work will be sprayed.
___ Pesticides should never be sprayed if they will contaminate the environment.

___ The department of agriculture should register pesticides.
___ More than one government agency should be involved in the registration of pesticides.

___ Only natural pesticides should be used because they are not dangerous to wildlife.
___ A new pesticide should be approved only if there is no other way to get rid of the pest.

___ All environmental problems are due to the misuse of pesticides.
___ The government should support research that would reduce pesticide use.

Name ______________________________ Date ________ Class Period ____

9. A substance called agent orange was sprayed on some of the forests in Vietnam. It contained a herbicide that caused the trees to lose their leaves. The herbicide was used so that the enemy could be spotted more easily.

Some of the Vietnam veterans filed a law suit because of health problems they claim are due to their exposure to agent orange. Would Rachel Carson have supported their right to file a law suit? ________ What evidence can you offer that supports your belief?

____

____

____

10. If Rachel Carson had been one of the authors of the Bill of Rights, what "right" would she have added? ____

____

11. Was the use of DDT banned before or after the publication of *Silent Spring*? _____

12. What new government agency was created after the publication of *Silent Spring* ?

____

Name ______________________ Date ________ Class Period ____

Unit III — 7.3 Pesticides — How Safe are They?

Introduction [277]

1. List three changes observed in mice that were exposed to pesticides during scientific studies. (1) ______________ (2) ______________ (3) ______________

2. List two problems caused by use of pesticides in the environment. (1) ______________ (2) ______________

Pesticide Poisoning [277]

3. **(True/False)** The use of pesticides has caused some agricultural workers to become ill and others to die.

4. List 3 reasons why farm workers might not report poisoning due to exposure to pesticides. (1) ______________ (2) ______________ (3) ______________

5. Explain why doctors might not recognize an illness as pesticide poisoning.
____
____

6. List three EPA regulations that protect farm workers from pesticide poisoning.
(1) ______________
(2) ______________
(3) ______________

7. What steps did the EPA take to protect farm workers from exposure to parathion.
____

8. How are most pesticides applied? ______________ What evidence indicated that crop dusters were affected by the exposure to the pesticides?
____
____

9. Describe situations in which pesticides might cause poisoning in people who are not farm workers.
____

The Banning of Pesticides [278]

10. Identify the government agency that controlled pesticide use:
    - Before 1970: ______________
    - Since 1970: ______________

11. The USDA promoted the use of pesticides to increase ______________.

12. Many of the first group of pesticides—the ______________ —were banned or their use restricted by the action of the EPA during the 1970s

Name ______________________________ **Date** ________ **Class Period** ____

13. Regulations for pesticide use made by the EPA are based on the results of ____________________, the effects upon ________________, or possible effects upon ________________.

14. The use of the herbicide 2,4, 5-T was restricted because it is contaminated with a very toxic substance called ________________.
The restrictions were based upon **(suspected/known)** effects upon:
( ) laboratory animals ( ) humans ( ) wildlife.

15. Most pesticides were registered **(before/after)** 1970. Were health and safety tests required for registration before 1970?_____

16. According to the National Academy of Science, we know the health effects of **(few/most/all)** of the pesticides we use.

17. Some ________ chemicals are used in pesticides. If only 25 of these chemicals can be tested each year, how many years will it take before scientists have sufficient data to determine health effects? ____________ How old will you be? ________

The Rules Have Changed [279]

18. New pesticides that are marketed or used in the U.S., it must be registered by the ______________________________.

19. Before it is registered the __________ and ________________ of its use must be evaluated.

20. Risks include how it might affect ______________ and the ________________.

21. Benefits evaluated include how it might protect crops and animals from __________ and ______________.

22. Identify four types of changes that scientists might observe in laboratory animals that are exposed to pesticides.
(1) ________________________ (3) ________________________
(2) ________________________ (4)________________________

23. Identify the four types of organisms that must be included when testing for the effects of pesticides. (1) ____________________ (2) ____________________
(3) __________________ (4) ____________________

Pesticides in Our Food [279]

24. What two types of information must the manufacturer include on a pesticide label?
(1) __________________________ (2) __________________________

25. Tolerance levels must be set for those pesticides that are used on ____________ __________. The **tolerance level** is the amount of pesticide that is considered ____________ when present in ____________ or ___________.

26. Identify the government agency responsible for:
• setting tolerance levels. ________________________________
• inspecting food for pesticide residue. ________________________________

Name ______________________________ Date __________ Class Period ____

27. **(True/False)** Since the FDA inspects food for pesticide levels, we can be certain that any food we buy does not contain unsafe levels of pesticides.

28. The phrase **days to harvest** refers to the number of days between the __________ ____________________ of the pesticide and the day the crop can be ____________ ____________________ Where can the grower find the information concerning the "days to harvest"? ______________________________

Pesticides in Our Water [280]

29. Some water supplies in agricultural areas have been found to contain fertilizers and __________________________.

30. Identify two areas where the pesticide BLADEX should not be applied.
(1) ______________________________
(2) ______________________________

31. What conditions allowed the pesticide Temik to reach the ground water in Suffolk County, Long Island? ______________________________
____

32. Temik was used by farmers on Long Island to reduce damage to the ____________ crop caused by ____________________.

33. What tolerance level has the state of New York set for Temik in drinking water? __________. How long will Temik remain in the wells? ____________________

34. Temik is biodegradable. Why will a biodegradable substance remain in the wells so long? ______________________________
____
____

How Safe is "Safe?" [280]

35. If scientists were asked to make a list of pesticides that are safe enough to be used by the typical homeowner, would they all agree? ________ If they were asked to determine the "tolerance level," would they all agree? ________

36. Why is the amount of pesticide allowed in food higher than the amount allowed in water? ______________________________
____

37. Tolerance levels are based on the diet of: ( ) **adults** ( ) **children**

38. The National Academy of Sciences suggests that children may be ingesting unsafe amount of pesticides. Give the two reasons why children may be at greater risk that adults.
(1) ______________________________
(2) ______________________________

39. Which of these actions is recommended by the National Academy of Sciences.
( ) **develop new test procedures to evaluate the toxicity of pesticides in children.**
( ) **reduce the amount of fruits and vegetable in children's diet.**

Name ______________________________ Date _________ Class Period ____

40. The manager of an orchard uses only registered pesticides, follows the directions for application of pesticides, and waits the correct number of "days to harvest" before picking the fruit. Pesticide residues remaining on the apples do not exceed the tolerance level set for any pesticide used. Yet there is some concern that the apples might not be safe. Explain.
____
____
____

Risk versus Benefits [281]

41. In order to enjoy the benefits of technology, we must also take certain __________.

42. Benefits can not be eliminated, but they can be ______________.

Reducing the Risks [282]

43. What risk was reduced by banning or restricting the use of chlorinated hydrocarbons? ______________________________

44. The risk to the consumer from the use of biodegradable pesticides is due to:
( ) **accumulation in fatty tissues**
( ) **toxic effects of breakdown products**

45. What requirement must be met before a person is certified and allowed to use pesticides that are classified as "restricted use"? ______________________________
____
____

46. Which of the following people can apply a pesticide that is classified as "restricted use"? [more than one may be checked]
( ) **homeowners** ( ) **all farmers** ( ) **certified commercial applicators** ( ) **farmers who have passed a written exam**

47. Why are workers required to wear protective clothing when applying certain pesticides. ______________________________
____

48. The "signal words" and "precautionary statements" required on pesticide labels are determined by the: ( ) **manufacturer** ( ) **EPA**

49. Which of the Toxicity Categories is the most toxic?
( ) **Category I** ( ) **Category II** ( ) **Category III**

50. Which of the following words is required on pesticides that are the most toxic?
( ) **CAUTION** ( ) **DANGER** ( ) **WARNING**

51. Which of the words in question 50 would most likely be found on pesticides applied by the typical homeowner? ____________________

52. The following directions, "Wash thoroughly with soap and water after handling", are found on pesticides in:
( ) **all categories** ( ) **category I** ( ) **category II** ( ) **category III**

Name ______________________________ **Date** ________ **Class Period** ____

53. List three risks that are increased when the amount of pesticide applied is greater than the concentration recommended on the label.
(1) ______________________________
(2) ______________________________
(3) ______________________________

54. Identify the weather condition that must be avoided when applying pesticides in order to prevent:
- damaging the plant: ______________________________
- excessive air pollution: ______________________________
- contamination of water supplies: ______________________________

55. Describe the best possible weather forecast for a farmer who is planning to apply a pesticide. ______________________________
____

56. Some plants, especially fruit trees, are sprayed with pesticides several times during the growing season. Which of the following spraying times would cause most damage to the honey bee colonies?
( ) **green tip** ( ) **pre-bloom** ( ) **full bloom** ( ) **post-bloom**

57. List three ways in which orchard owners, who must spray plants when they are blooming, can reduce the number of bees affected by the spray.
(1) ______________________________
(2) ______________________________
(3) ______________________________

58. Give two advantages of the **microencapsulation** of pesticides such as parathion.
(1) ______________________________
____
(2) ______________________________

59. How does microencapsulation of pesticides make them more harmful to bees?
____

How can the farmer reduce the risk for the bees? ______________________________
____

The Perfect Apple [284]

60. Explain the statement: "You are one reason why farmers use so many pesticides."
____

61. Would you be willing to use oranges with "rust spots," if the damage was only skin deep and the fruit was not affected? __________
Would you consider eating sweet corn that had minor insect damage? [Assume that the damage could be removed by cutting off the tip of the ear of corn]? __________

62. Assume that you are given a choice of a salad made with fresh pineapple or one made with canned pineapple. If you knew that the fresh pineapple had been dipped in a fungicide solution to prevent fresh fruit rot and the canned pineapple had not been treated, which would you select? __________ Explain why. __________

Name ______________________________ **Date** ________ **Class Period** ____

63. Do you think food should have labels which list the pesticides that have been used? __________ If you were given the choice of buying food with a label at a higher price than food that was not labeled, which would you select? __________

64. Do you think it would make a difference if consumers refused to buy frozen orange juice that was made from oranges treated for damage caused by thrips or other insects that does not affect the fruit quality? __________

Is There a Better Way? [284]

65. Are the pesticides in the environment today harmful to human health? __________ Are you sure? _________ Explain. ______________________________
____
____

66. Most testing of pesticides has been based on the:
( ) **short-term effects** ( ) **long-term effects** from exposure to:
( ) **small amounts** ( ) **large amounts of pesticides**

67. Do you think that scientists can really determine "safe" levels for exposure to pesticides? __________ Explain. ______________________________
____
____

68. Do you think that the amount of pesticides used should be reduced? __________
Would you be willing to pay more for food grown without pesticides? __________
Would you be willing to use food that was not as perfect if it meant that fewer pesticides would be needed to produce it? __________

Name ______________________________ Date ________ Class Period ____

Unit III — 7.4 IPM — A Better Way to Control Pests

Introduction [287]

1. Pests normally destroy ________________ of our food each year.

2. Circle those foods that would probably disappear from most grocery stores if pesticides weren't used.

apples	**bananas**	**peaches**	**strawberries**
lettuce	**green beans**	**peanuts**	**tomatoes**

3. Give one example of evidence that tell us pesticides are not eliminating the pests.
____
____

4. **IPM** or **I**________________ **P**__________ **M**________________ uses a ________________ of methods to control insects. The goal of IPM is to ________________ not eliminate the population of a particular pest.

Resistant Varieties [287]

5. Two types of plants that produce their own pesticides are: ________________ and ________________

6. Identify three chemicals produced by plants that act as natural pesticides.
(1) ________________ (2) ________________ (3) ________________

7. Identify the two body systems in bugs that are affected by pesticides.
(1) ________________ (2) ________________

8. Identify two physical barriers that protect some plants from pests.
(1) ________________ (2) ________________

9. Resistant varieties of each of the following crops have been developed. Identify the pest or pests that attack varieties of these crops that are not resistant.
wheat—________________ **alfalfa**—________________

10. Often when plants are in storage they are attacked by insects or fungi. What is one way we could avoid loss of food crops [during storage] without using pesticides?
____

11. Give two reasons why scientists have not been able to develop more varieties of crops that are resistant to pests. ________________________________

12. Scientists have developed apple trees that are resistant to four major diseases (scab, rust, fire blight, and mildew). How many years did this process take? _________

13. Plant breeders are trying to develop a peach tree that is resistant to the peach borer. What other characteristics must this peach tree have in order to be acceptable to the consumers? ________________________________

Name ______________________________ Date ________ Class Period ____

14. Using **genetic engineering** techniques, scientist may be able to transfer the __________ or chemical message for the synthesis of nicotine from tobacco plants to potato plants. The ability to transfer genes will reduce the ___________ needed to develop a plant variety that is resistant to a certain pest or disease.

Beneficial Insects [288]

15. What type of insect is the most beneficial insect for controlling populations of pests? ____________ The females deposit their eggs inside the __________ or ____________ of the insect pest. The pest provides ____________ for the parasitic wasp larvae.

16. List four insect pests that are controlled by parasitic wasps.
(1) _________________________ (3) _________________________
(2) _________________________ (4) _________________________

17. List three kinds of insects that are good biological controls for aphids.
(1) _______________ (2) _________________ (3) ____________________

18. Why isn't the praying mantis a good choice for insect control?
____

Other Natural Enemies [289]

19. Identify the type of organism listed below, and give the pest(s) that it helps control.

Some fungi in the soil help control harmful __________________.
Nosema locustae a one-celled _____________ that helps control ____________
Bacillus thuringiensis (Bt) is a _______________ that helps control ___________
A new variety of Bt known as BTI may help control larvae of the ______________
by affecting their ________________ system.
Bacillus popillae is a _________________ that causes milky spore disease in the larvae of the ___________________________.

20. Bt is a very ________ insecticide because the toxin it produces does not harm beneficial organisms and is not harmful to humans. Bt is also a very ________________ insecticide because it only affects ________________.

21. How many days must the grower wait before harvesting a vegetable that has been sprayed with Bt? _________ Does the FDA inspect vegetables to determine the levels of Bt present? ____
____

22. What are two disadvantages of using milky spore disease to control the Japanese beetle? (1) _____________________ (2) ___________________________
____

23. Identify two types of animals, other than insects, that are effective predators of some pests. (1) ____________ (2) ____________

Name ______________________________ **Date** ________ **Class Period** ____

Birth Control [290]

24. Radiation is used to sterilize **(female/male)** insects that are released into the environment. This reduces the population of insects by reducing the number of eggs that ___________.

25. Radiation was used to eliminate the population of ____________________ introduced into Florida.
26. Sterilization is an effective means of control when the population is ___________ and not ____________________.

27. Do you think that sterilization would be an effective way to control the Africanized bee? _________

Creating Confusion With Chemicals [290]

28. Which of the following insect populations has been successfully controlled by using sex attractants? ( ) **gypsy moth** ( ) **Mediterranean fruit fly**

29. Some California farmers are using the sex attractant of the female _____________ in an attempt to decrease the amount of pesticides used on the ____________ crop.

30. The chemicals that are placed in traps to catch insects are: ( ) **synthetic growth hormones** ( ) **sex attractants**

31. Describe changes caused by synthetic hormones that could reduce the number of adult insect pests. ____

____

32. Which of the following statements is true?
( ) Some synthetic hormones that control development of insects are toxic to humans.
( ) Some synthetic hormones that control development of insects interfere with the development of beneficial species.

Cultivation Practices [290]

33. List three cultivation practices that help reduce the population of insect pests and describe how each one reduces the pest population.
(1) ____
(2) ____
(3) ____

Pesticides [291]

34. Biological controls have:
( ) reduced the use of pesticides ( ) eliminated the use of pesticides

35. Which state has significantly reduced the amount of pesticides used to control mosquitoes? __________________
Which state has significantly reduced the amount of pesticides used to control pests on the cotton crop? ______________

Name ______________________ Date __________ Class Period ____

36. A computer program at Texas A and M University tells farmers when it is necessary to spray their cotton crop. What information must be fed into the computer?
____
____

37. What information must be fed into the computer program that is designed to help the orchard owner control the population of codling moths in the apple orchard?
____

38. The development of insects is dependent upon ______________. Thus the farmer needs to know the number of ______________ ______________ days. to know when the insects at the correct stage of development to spray.

The Benefits and Disadvantages of Biological Controls [292]

39. List three advantages of biological controls.
(1) ____
(2) ____
(3) ____

40. Why are biological controls often more expensive to develop than chemical controls? ____
____

41. Why aren't more private companies interested in developing biological controls?
____
____

42. Are biological controls available for most species of pests? ________________
____

43. Why is it sometimes easier to use chemical pesticides rather than a biological control? ____
____

44. What is the major disadvantage of milky spore disease? ________________
____

43. Why is it so difficult to convince some people that biological controls really work?
____

Pest Control Does Not Mean No Pests [292]

44. A pest is only a pest when ____

45. The goal of IPM is to **(eliminate/reduce)** the pest population.

46. Farmers will probably accept an IPM plan if they can be convinced that the amount of money they will lose because of damage done by the pest is less than _______
____

47. While a large population of *Lygus* bugs **(decreases/increases)** the yield, a small population will **(decrease/increase)** the yield by ________________
____

Name ______________________________ Date ________ Class Period ____

Unit III — 7.5 The Banning of Pesticides

The Law is Changed [296]

1. The type of tests required by the 1947 Federal Insecticide, Fungicide, and Rodenticide Act were **(short-term/long term).**

2. The amendment added in 1972 required tests for **(short-term/long-term)** or **(acute/chronic)** effects.

3. The first pesticide to be banned was ________. It was banned because of its effects upon **(humans/wildlife)** and because it caused ______________ in some species.

4. Have most of the chemicals in pesticides sold today been tested for long-term effects? ________ How long will it take for all the chemicals in pesticides on the market to be tested? ___________

5. Were you aware that many of the pesticides that you buy have not been tested for long-term effects? ________

A Partial List of Pesticides that Have Been Banned [296]

6. List 3 reasons why certain pesticides were banned.
   (1) ____________________________ (2) ____________________________
   (3) ____________________________

7. Identify the organ affected by chlordane and heptachlor. ________________

8. Why were exemptions allowed for chlordane and heptachlor? ________________
   ____

9. Which pesticide was banned because it contaminated the Great Lakes? __________

10. Which pesticide was banned after it contaminated the St. James River? _________

11. Which pesticide caused severe disorders of the nervous system? _____________

12. Which pesticide was linked to genetic mutations? ___________

13. Which pesticide caused sterility among workers? ___________

14. Which pesticide was found in the drinking water in the San Joaquin Valley ?
    ____

15. Sale and use of _________________ was immediately suspended because of the high risk of birth defects.

16. ____________________ was the first pesticide banned by a state because of its effects on wildlife.

Name ______________________________ **Date** ________ **Class Period** ____

17. If the sale of a pesticide is banned in the United States, do you think the production of the pesticide should be banned? ______________ Justify your position. ____

A Exemption Caused Trouble in Hawaii [297]

18. Which pesticide caused problems in Hawaii? __________________

19. The pesticide was sprayed to control the _______ that were protecting ________ ________ that damage the ____________________ crop.

20. Why didn't the growers use natural predators of the mealy bugs instead of the pesticide? ______________________________________

21. Why did the EPA allow the growers to use a pesticide that had been banned? ____

22. Draw a food chain to show how heptachlor contaminated the supply of milk and other dairy products.

23. The tolerance level set for heptachlor in food was _______ ppm. The level found in milk was _________ ppm.

24. We may never know if any health problem are caused by heptachlor. Why? ____ ____ ____

A Costly Mistake in the Midwest [298]

25. The pesticide found in milk in the Midwest was __________________.

26. The contaminated grain was used to make _______________ which was used in the production of __________________.

27. The part of the grain, called ____________, that remains after alcohol is produced is used as a feed supplement.

28. Draw a food chain that shows why the farmers were concerned.

29. Were there any short-term health effects? ______

30. Which humans were at the greatest risk of damage from the pesticide in the milk? _______________ Why? ______________________________ ____

Name ______________________________ Date ________ Class Period ____

Keeping the Grass Green [299]

31. Which pesticide has caused the death of ducks and geese on golf courses?
____

32. Do you think that lawn-care companies should be required to post signs that indicate a lawn has been sprayed? __________
Why or why not? ______________________________

33. Some lawn-care companies are investigating alternatives to pesticides. How might they control pests without using pesticides?
____

Made in the U.S.A. [299]

34. Chemical companies can sell pesticides that have been banned or restricted, and pesticides that have not been approved by the EPA. Who buys them?
____

35. List two ways that government regulations try to protect people in countries that buy the pesticides.
(1) ______________________________
(2) ______________________________

36. Give two reasons why the exported pesticides are often the cause of pesticide poisoning. (1) ______________________________
(2) ______________________________

37. Why do farmers in other countries use pesticides like DDT when safer pesticides are available? ______________________________

38. Why aren't people in third world countries concerned about the long-term effects of pesticides? ______________________________
____

39. Are you concerned about other countries using pesticides that are banned in the United States? __________

40. How might pesticides that are used in other countries affect you and your environment? ______________________________
____

41. What agency is responsible for monitoring the pesticide levels in food? ________
Do you think that one sample for 14 million pounds of oranges is enough? _______
Why aren't more samples taken? __________________
Should Congress approve more money for the FDA? ________________

Name ______________________________ Date ________ Class Period ____

UNIT IV — 1.1 Aquatic Environments

[page 305]

1. Aquatic ecosystems include __________ water ecosystems and ________ water ecosystems.

2. Freshwater ecosystems include ecosystems with ____________ water and ecosystems with _____________ water.

3. The characteristic that is used to classify a body of water as a lake is **(depth/surface area/volume)**.

4. The term that can <u>not</u> be used to describe a shallow body of flowing water is **(brook/creek/river/stream)**.

5. Streams, brooks, creeks or smaller rivers that empty into a larger river are called the ____________________ of the river.

6. Using a map of your state, locate the major rivers that drain water from the land. In the space below, draw an outline of the state and show the location of these rivers.

7. Locate the city or town where you live on the map above. Trace the path of a drop of water that runs off the football field at your school by listing the names of the tributaries and the river(s) that carry the water to the ocean.

____

____

8. By studying a map, you may be able to distinguish human-made channels from natural rivers or streams. How do human-made channels differ from naturally flowing bodies of water? ______________________________________

9. Study the map of your state and try to identify rivers or streams that have been channelized. List any examples you find below.

____

____

Name ______________________ Date __________ Class Period ____

10. What is the advantage of changing a natural stream into a straight box-like channel?

____

11. What are the disadvantages of changing a natural stream into a human-made channel?

____

12. Match the organisms below with the aquatic habitat that they require.

(1) bottom (2) surface film (3) open water (4) water's edge

_____ A. the mosquito larvae—an air-breathing insect.
_____ B. microscopic plants and animals that are floaters—plankton.
_____ C. large fish.
_____ D. bacteria, snails, crayfish, and other organisms that feed on decaying matter.
_____ E. small fish.

13. Which habitat provides homes for the greatest number of species? _____________

14. Match the stream habitat listed below with the correct descriptions.

(1) lower reaches (2) middle reaches (3) upper reaches

_____ A. greatest slope.
_____ B. gentle, steady flow of water
_____ C. bottom is rocky, due to constant erosion.
_____ D. least slope.
_____ E. organisms are most like those in a pond.
_____ F. while some areas in this region are eroded others are covered with silt.
_____ G. large, flat floodplains border this region.
_____ H. waterfalls.
_____ I. fewer species and smaller populations than other regions.
_____ J. many riffles and rapids.
_____ K. bottom is covered with sand or mud.
_____ L. pools are large, some riffles and rapids.

Name ______________________ Date ________ Class Period ____

UNIT IV — 1.2 Creatures that Live in the Water

[page 308]

1. The best indicators of water quality are:
   ____ amphibians ____ fish ____ aquatic insects ____ mammals

2. In addition to chemical tests, a study of water quality often includes a survey of living organisms. Explain why living organisms are sometimes a better indicator of water quality than chemical tests?
   ____
   ____
   ____

3. Organisms whose presence in large numbers indicates the quality of water are sometimes called **indicator species** or __________ **species.**

4. Identify the habitat(s) where you would expect to find each organism listed below. Read the background information provided for each organism and then indicate where you would most likely find the organism. **Two or three habitats (letters) may be listed for some organisms.**

   B = bottom E = water's edge S = surface O = open water

________	*Tubifex* worm	________	water striders
________	leeches	________	giant water bugs
________	flatworms	________	diving beetles
________	crayfish	________	whirligig beetles-adult
________	scuds	________	whirligig beetles-larvae
________	water fleas	________	scavenger beetles
________	copepods	________	dragonfly nymphs
________	seed shrimps	________	damselfly nymphs
________	spiders	________	mayfly nymphs
________	water mites	________	stonefly nymphs
________	mosquito larvae	________	caddisfly nymphs
________	midgeflies	________	dobsonfly larvae
________	craneflies	________	springtails
________	backswimmer	________	water pennies-adult
________	water boatman	________	water pennies-larvae
________	water scorpions	________	snails
________		________	mussels-adults

Which habitat contains the most species? ______________________

Why are more species found in this habitat? ______________________
____

Which habitat contains the fewest species? ______________________

Why aren't more species found in this habitat? ______________________

Name ______________________________ Date __________ Class Period ____

5. Read the "environmental factors" section for each of the following organisms, and identify the water quality that is indicated by the presence of large numbers of the organism. Choose the best answer.

_____ A. Tubifex worms
_____ B. scuds
_____ C. water fleas
_____ D. mayfly nymphs
_____ E. stonefly nymphs

(1) clean water
(2) clean water/high oxygen
(3) organic pollution/low oxygen

6. Using the letters indicated below. Identify the niche of each of the following organisms. Some groups include organisms that are in two or more niches.

H = herbivore **O = omnivore** **S = scavenger**
P = predator **Pa = parasite**

__________ *Tubifex* worm
__________ leeches
__________ flatworms
__________ crayfish
__________ scuds
__________ water fleas
__________ copepods
__________ seed shrimps
__________ spiders
__________ water mites
__________ mosquito larvae
__________ midgeflies
__________ craneflies
__________ water boatman
__________ water scorpions
__________ backswimmers
__________ water striders
__________ giant water bugs
__________ diving beetles
__________ whirligig beetles-adult
__________ scavenger beetles-adult
__________ scavenger beetles-larvae
__________ dragonfly nymphs
__________ damselfly nymphs
__________ mayfly nymphs
__________ stonefly nymphs
__________ caddisfly nymphs
__________ dobsonfly larvae
__________ springtails
__________ water pennies-larvae & adult
__________ snails
__________ mussels-adult

7. Match each of the following organisms with the appropriate description(s).

_____ A. host for parasite that causes malaria
_____ B. host for lung fluke
_____ C. some species are parasites on mosquitoes
_____ D. carry viruses that cause encephalitis
_____ E. larvae are parasites on fish
_____ F. host for larvae of water mites

(1) crayfish
(2) mites
(3) mosquitoes
(4) mussels
(5) water strider

8. Identify the structures used for "breathing" by each of the following organisms.

_____ A. mayfly nymph
_____ B. water scorpion
_____ D. mosquito larvae
_____ E. diving beetle
_____ F. spiders
_____ G. scavenger beetles

(1) body hairs trap layer of air
(2) air tubes
(3) club-shaped antennae
(4) opening in abdomen
(5) seven pairs of gill filaments

Name ______________________________ Date __________ Class Period ____

9. Identify each of the organisms described below.

_____ A. also known as hellgrammite	**(1)**	**diving beetle**
_____ B. also known as sideswimmer	**(2)**	**dobsonfly larvae**
_____ C. pest at fish hatcheries	**(3)**	**caddisfly larvae**
_____ D. larvae are called water tiger	**(4)**	**crayfish**
_____ E. larvae look like a centipede	**(5)**	**giant water bug**
_____ F. live in cases made of mud or stones	**(6)**	**scud**
_____ G. pest that feeds on farmer's crops	**(7)**	**whirligig beetle**

10. For each of the organisms listed below, indicate whether they are important as food for fish **F,** they are a predator that eats fish **P,** or they are a parasite on fish **Pa**. If they eat small fish and are eaten by larger fish use **P F**. If they do not eat fish and are not eaten by fish use **N**.

_____ *Tubifex* worm	_____ midgeflies	_____ dragonfly nymphs
_____ leeches	_____ craneflies	_____ damselfly nymphs
_____ flatworms	_____ water boatman	_____ mayfly nymphs
_____ crayfish	_____ backswimmers	_____ stonefly nymphs
_____ scuds	_____ water scorpions	_____ caddisfly larvae
_____ water fleas	_____ water striders	_____ dobsonfly larvae
_____ copepods	_____ giant water bugs	_____ springtails
_____ seed shrimps	_____ diving beetles	_____ water pennies
_____ water mites	_____ whirligig beetles	_____ snails
_____ spiders	_____ scavenger beetles	_____ mussels
_____ mosquito larvae		

Name ______________________________ Date __________ Class Period ____

UIV — 1.3 Plants in an Aquatic Ecosystem

Introduction [319]

1. Underline the statement that an ecologist would consider the most important.
   A. Algae is a protist. B. Algae is not a plant. C. Algae is a producer.

Types of Algae [319]

2. Match the statements below with the type of algae described.

   (1) filamentous algae (2) stonewort algae (3) plankton algae

_____ A. cells are arranged in long strands.
_____ B. microscopic, floating alga cells.
_____ C. pond scum.
_____ D. grows on the bottom of lakes in limestone soil.
_____ E. fur-like covering on stones.
_____ F. colors the water when present in large numbers.
_____ G. a population explosion of this algae is called an "algal bloom."
_____ H. floating mats of algae.
_____ I. may grow to two or three feet in length with a central stem that has whorls of branches.

3. Since algae are not flowering plants, and **algal bloom** does not refer to the presence of many flowers, but to the presence of many alga __________. An algae bloom often makes the water appear pea-soup _________, but it may also be __________, ____________ or ______________.

The Importance of Algae [319]

4. List two reasons why algae are important in an aquatic ecosystem.
   (1) ____
   (2) ____

5. Algae produce oxygen during ______________________. Both fish and algae need oxygen for ____________________. Algae die if they can not get enough ____________. As more algae die, the need for oxygen **(decreases/increases)**.

6. The level of dissolved oxygen may be dangerously low after a series of **(cloudy/sunny)** days when the temperatures are **(cool/warm)**.

7. List four possible problems that might be caused by an **algal bloom**.
   (1) __________________ (2) __________________
   (3) __________________ (4) __________________

8. Water treatment plants remove algae by filtering the water through a __________ filter. When the filter becomes clogged, it is cleaned by a process called a _________________. This process involves stirring the sand and ___________ the direction the water moves.

9. The length of time between backwashes is reduced during the ____________ months because of the increased number of algal blooms.

Name ______________________ Date __________ Class Period ____

Types of Flowering Plants [320]

10. Identify each type of flowering plant described below.

_____________ grow in shallow water—most of the plant is above the water's surface.
_____________ grow in deep water—only the flower rises above the water's surface.
_____________ The plant is rooted in the muddy bottom—only the leaves and flowers are slightly above the surface of the water.

Importance of Flowering Plants [321]

11. Briefly explain how flowering plants are important to each of the following organisms.
(1) bacteria and insect larvae: ______________________
(2) fish and frogs: ______________________
(3) waterfowl: ______________________
(4) muskrats: ______________________

12. The emerging plants growing along the water's edge provide ________ and __________ for wildlife and prevent ______________.

13. A temporary pond may be a better habitat for young frogs (tadpoles) than a larger pond because there are no ___________.

Water Pollution and Plant Growth [321]

14. Primary nutrients that stimulate plant growth include: ________________ and ____________________.

15. Identify pollutants that are classified as:.
point sources: ______________________
nonpoint sources: ______________________

16. Name three different types of waste that contain nutrients that stimulate plant growth.
(1) _______________ (2) ______________________ (3) ___________________

Water Use and Plant Growth [322]

17. Identify three uses of water that might require control of plant growth.
(1) ______________________
(2) ______________________
(3) ______________________

18. List two reasons why it is desirable to have some aquatic plants.
(1) ______________________
(2) ______________________

Controlling Plant Growth [322]

19. To prevent excessive growth of submerged plants, the minimum depth of a pond should be _________ feet or ________ meters.

Name ________________________________ Date __________ Class Period ____

20. Physical removal of plants or the use of animals that eat plants may not be an effective way to decrease plant growth because new plants grow from __________
____

21. Why do some states prohibit the release of species of fish that might control plant growth? ____

22. The most effective way to control plant growth is the use of __________________. The first step in the treatment process is always to ____________________
____

23. The best time for treatment is usually **(early spring/early summer/late fall).**

24. What information does the person applying the chemical need to know in order to determine the proper amount of chemical to use? ____________________
____

25. It is necessary to obtain a permit to chemically treat a pond or lake in your state?
____

26. The owner of a pond identifies the pond weeds, obtains the proper permit, buys the proper herbicide, and applies it according to the directions given by the manufacturer. The label says that the herbicide "will not harm fish" if applied according to directions. Yet a few days later there is a "fish kill." Give two possible reasons why the fish died.
(1) ____
(2) ____.

Name ______________________ Date ________ Class Period ____

Unit IV — 1.4 Water Quality — Chemical and Physical Factors

Introduction [325]

1. When judging the quality of apples or water, the _________ is an important factor that must be considered.

2. Identify three factors or conditions that must be examined to determine water quality. (1) ______________ (2) ______________ (3) ______________

3. Identify a chemical that is considered a pollutant in high quality drinking water, but is sometimes a pollutant in a stream ecosystem or an aquarium. ______________

4. A substance **pollutes** the water if it is present in ______________________
____

5. When a number of different chemicals are added to the water, we must consider the toxic effects of each chemical alone, and also the ______________ effects that occur when more than one chemical is present.

Dissolved Oxygen [325]

6. Check the name of the chemical that is the source of oxygen for aquatic plants and animals.
____ water molecules (H_2O) ____ oxygen (O_2) dissolved in the water

7. The molecules of oxygen that are dissolved in the water come from the ______________ or the process of ______________.

8. For each of the following situations, place a checkmark beside the situation in each pair that you would expect to have the highest level of dissolved oxygen.

A.		water flowing rapidly in a stream.		water standing in a swamp.
B.		a pond on a sunny day.		a pond at night or after a series of cloudy days.
C.		deep water in a lake.		water near the surface of a lake.
D.		water with a high level of organic matter.		water with a low level of organic matter.
E.		water with a large variety of organisms.		water with only a few types of organisms.
F.		water with mostly air breathing invertebrates and carp.		water with stoneflies and trout.

9. Explain why the level of oxygen is low in each of the following situations:

   A. Water that is polluted with organic matter has a low level of dissolved oxygen because the ______________________________

   B. A pond with a large algae population has a low level of dissolved oxygen early in the morning because ______________________________
   ____

Name ______________________________ Date ________ Class Period ____

10. A biological survey of a stream reveals that it contains several species of fish and a large variety of macroinvertebrates. The minimum amount of DO that you would expect to find is ______ ppm.

Temperature and Dissolved Oxygen [326]

11. **Solubility** refers to the ability of a substance to ______________ in a liquid.

12. As the temperature of a liquid increases the amount of gas that can be dissolved in the liquid **(decreases/increases)**.

13. What is the maximum temperature of the water in which a fish could live if the fish required 10 ppm of dissolved oxygen? ______________

14. **(Carp/Catfish/Trout)** would probably not be found in a reservoir that receives water from an electrical power plant.

15. Fish are **(cold/warm)** blooded. As the temperature of the water increases their body temperature **(decreases/does not change/increases)**.

16. As the temperature of the water increases, the amount of oxygen needed by the fish **(decreases/does not change/increases)**.

17. Give two reasons why fish may not be able to get as much oxygen as they need if the water is warm.
(1) ____
(2) ____

18. To prevent increasing the temperature of a river or stream beyond acceptable limits, industries send the hot water to a ____________________ or a ____________ ________________.

Can There Be Too Much Oxygen? [327]

19. A liquid holding all of a substance that can be dissolved at a given temperature is ______________________.

20. A liquid holding more of a substance than is normally dissolved at that temperature is ______________________.

21. When fish get too much air in their blood, the gases form ______________ and the condition is called ______________.

22. Fish most often affected by this disease live **(upstream/downstream)** from a dam. The condition most often occurs during the ______________ season when water levels are high..

23. A healthy aquatic ecosystem normally has water that is more than _______ percent saturated but less than _______ percent saturated.

Name ______________________________________ **Date** _________ **Class Period** ____

24. Give two possible reasons why water might be less than 60% saturated with oxygen.
(1) ____
(2) ____

pH [327]

25. Most organisms live in water with a pH between _____ and _____.

26. The pH of a lake is 5. Would you expect to find a large population of algae? _____

27. The pH of a lake is 10. Would you expect to find a large variety of fish? _____

28. The pH of water that will support a large variety of plant and animals life is between ______ and ______.

29. New ponds and lakes usually have a pH **(below 7/above 7)**. As lakes age, the pH usually **(decreases/increases)**. This change in pH is due to the production of ______________________.

30. Chemicals, such as detergents, sometimes cause fish kills by altering the water's ________.

31. Some coal deposits contain __________ that forms an acid when mixed with water. Lakes formed by the strip mining of coal do not contain fish because the water is too ____________. When it rains and the lakes overflow, the change in pH causes a ________ __________.

Hardness [328]

32. The hardness of a body of water is determined by the type of ___________ and __________ in the watershed.

33. Lakes that are located in areas with granite rock contain **(few minerals/many minerals)** and the water is **(hard/soft)**.

34. Bodies of water in areas with limestone rocks are **(hard/soft)** due to the presence of the minerals _______________ and ___________________.

35. Total hardness is the test most frequently used by industries because certain ________________ interfere with industrial processes.

36. Hard water is **(less/more)** affected by acid rain because the ________________ act as buffers to prevent changes in the ________.

Nitrates and Nitrites [329]

37. Fertilizers used on lawns, golf courses and farm fields contain _______________.

38. Other major sources of this nutrient are ___________________________________.

39. Levels of nitrogen greater than 0.30 ppm may cause an ____________________.

Name ____ Date __________ Class Period ____

40. **(Adults/Babies)** are affected by drinking water with high levels of nitrates. The condition is most likely to occur **(in cities/on farms)**.

41. Babies with the condition **methemoglobinemia** are a ________ color because the **<u>nitrites</u>** interfere with the ability of hemoglobin to __________________________.

42. Well water should be checked for nitrates in areas where there are __________ or homes with ________________________. Can you think of other areas where nitrate contamination of well water might occur? ______________________________

Phosphates [329]

43. Phosphates enter water from (1) ______________________ (2) ________________, and (3) ______________________________.

44. The major problem caused by excess phosphates is ____.

VOCABULARY REVIEW

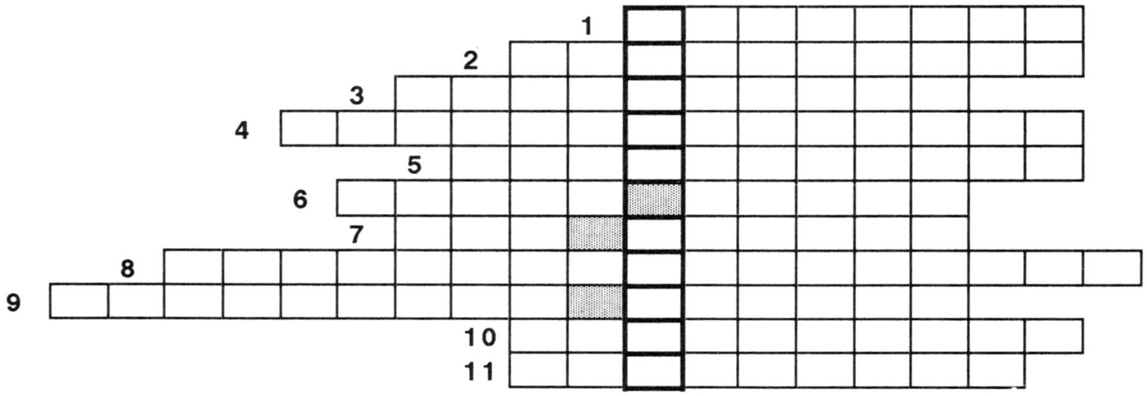

1 describes a lake that has a pH of 8.0.
2 the ability of a substance to dissolve in a liquid.
3 chemical in red blood cells that carries oxygen.
4 condition in which more gas is dissolved in water than would normally be dissolved at the given temperature.
5 The poisonous effect of a chemical is greater in the presence of other chemicals.
6 occurs when the water contains more than 0.30 ppm nitrogen.
7 condition that occurs in fish when the water is supersaturated.
8 occurs in babies when the water they drink contains excess nitrates.
9 is often a limiting factor in ponds but not in rapidly flowing streams.
10 nutrient that may contaminate water and cause algal blooms.
11 carbonic acid in rain often dissolves calcium and magnesium from these rocks.

Name ______________________________ Date __________ Class Period ____

Unit IV — 2.1 Changing the Flow — What it Does to Wildlife

Nature's Engineers [331]

1. The animal that is a natural engineer is the ____________.

2. The trees used to build dams include __________, __________, and __________.

3. To get trees to the building site, beavers sometimes dig ____________.

4. The activities of beavers help conserve ____________ and __________.

5. The beaver ponds help increase the supply of ____________ water by allowing more time for ______________________.

6. Beaver left southwestern Wyoming because the trees had been destroyed by the ____________. The beaver need trees for ____________ and ______________.

7. The amount of silt carried by the water is greater in a stream **(with/without)** beaver dams.

8. Give two reasons why hiring government employees to provide beaver with materials to build dams is a better approach than having the employees build rock chip dams. (1) ____
(2) ____

9. The beaver dams help reduce ____________ and improve the ____________ for many other organisms.

10. Explain how the beaver's actions increase the productivity of the stream.
____
____

11. Streams with beaver dams provide better nurseries for young fish than free-flowing streams because of the increased amount of ____________ and ____________.

12. Why are beavers moving to suburban areas? ______________________________
____

Danger Below the Dam [332]

13. The worst fish kill recorded in the state of Missouri was due to excessive amounts of **(acid/air)** dissolved in the water.

14. Explain how water becomes supersaturated. ______________________________
____
____

15. The amount of air dissolved in water can be harmful to wildlife if it is more than _____ percent above normal.

16. If the water is supersaturated the fish may develop ________________ disease.

Name ______________________________ Date __________ Class Period ____

17. Identify two ways in which gas bubble disease affects the eyes of fish.
(1) ____
____
(2) ____

18. Explain how gas bubble disease can cause the death of a fish.
____

Barriers for Fish [333]

19. In addition to supersaturation, give two ways that dams reduce the fish population.
(1) ____
(2) ____

20. Describe a solution for each problem listed in question 1 and give any limitations or disadvantages associated with each solution.
(1) ____
____
(2) ____
____

21. Just like any other construction project, the building of a dam has both environmental costs and benefits.

Give two ways in which the environmental changes created by a dam may reduce the income of people living in the area.
(1) ____
(2) ____

Give two ways that a dam might improve the local economy .
(1) ____
(2) ____

22. Do you think that dams which are no longer serving their original purpose should be removed to help the migration of fish? __________
If so, who do you think should pay for their removal. __________

23. If you were in charge of issuing permits for small hydroelectric dams on the Idaho River, how many of the 50 proposed dams would you approve? __________
Why this number? ____

Channelized Streams and Wildlife [334]

24. The path of a stream or river, where water normally flows, is called the river bed or ____________________.

25. The channels of old rivers are like **(interstate highways/winding country roads).**

26. Give two reasons why rivers are deepened and straightened.
(1) ____
(2) ____

Name ____ Date __________ Class Period _____

27. Indicate whether channeling a stream will **I** increase or **D** decrease each of the following:

____ water temperature	___ rate of erosion	___ food supply	___ productivity
____ dissolved oxygen	___ breeding places	___ turbidity	

28. Fish are not the only form of wildlife affected by the straightening of the river's channels. Identify three other species of animals that are affected by channeling a stream. ____

The Kissimmee — Repairing a River [335]

29. Why did the Army Corps of Engineers straighten the Kissimmee River?
____
____

30. Now the Army Corps of Engineers is putting the bends back in the river. Explain how this will help wildlife: ____
people: ____

What Have We Learned? [335]

31. Changing the natural flow of water affects both the quality and quantity of ______________ for wildlife and ______________ for the people.

Name ______________________ Date __________ Class Period ____

Unit IV — 2.2 Trouble at Mono Lake

Introduction [337]

1. Explain why Mono Basin has a desert climate. (Hint: see Unit II 2.1 Weather Patterns) ______________________
____

The Inland Sea [337]

2. List the three sources of the water in Mono Lake. (1) ______________ (2) ______________ and (3) ______________

3. How is Mono Lake like the ocean? ______________________
____

4. If you found yourself at Mono lake without anything to drink, could you drink the water? ______________________

Life in Mono Lake [337]

5. The lake is sometimes referred to as the Dead Sea of California. Is this description appropriate?_____ Explain ______________________

6. Draw a food web that includes only the producers and consumers found in and near the lake. As you continue to read add other organisms.

7. What commercial product is harvested from the lake? ______________ Draw the food chain which includes this product.

Prepare for Take Off [337]

8. Mono Lake is an important **(feeding station/nesting area)** for the Wilson's phalarope.

9. Describe two important changes that occur during the phalarope's stay at Mono Lake. (1) ______________________
(2) ______________________

10. Give the location where the Wilson's phalarope can be found:
during the summer: ______________________
during the winter: ______________________

Name ________________________________ **Date** ________ **Class Period** ____

Don't Step on the Eggs [338]

11. What bird species nests at Mono Lake? ______________________________

12. List three factors that made Mono Lake an ideal habitat for the nesting birds.
(1) __________________________ (2) __________________________
(3) ____

13. What change in the lake has made the habitat less suitable for the nesting birds?
____
____

14. What is the major predator of the sea gulls? ______________________

Tourists at Mono Lake — Then and Now [338]

15. The size of Mono Lake has **(decreased/increased)** since the 1950s.

16. Explain how the changes have made it unpleasant and sometimes dangerous to:
swim or boat: ____
breathe the air: ____

17. ____________________ are columns of calcium carbonate that form when ________________ reacts with chemicals in the lake.

Where Did the Water Go? [339]

18. What city gets part of its water supply from the Mono Basin? ______________

19. The city obtained the water rights for the Mono Basin from the ______________ ________________ or the ______________________________.

20. List the four sources of water used by Los Angeles beginning with the major source. Rank the sources from most important to least important.
(1) __________________________ (2) __________________________
(3) __________________________ (4) __________________________

A Day in Court [339]

21. The major land owner in the Mono Basin is the ________________________

22. Do you think that the Los Angeles Department of Water and Power should have the right to divert water from Mono Basin if it will harm the lake? ______________

23. According to the 1983 ruling made by the California Supreme Court, what must be considered when granting water rights? ______________________________
____

The Need for Water [340]

24. The federal task force recommended that Los Angeles:
___ stop diverting water from the Mono Basin.
___ reduce the amount of water the city diverts from Mono Basin.
___ continue to divert water from Mono Basin to meet the city's needs.

Name ______________________ Date ________ Class Period ____

25. List the two recommendations made by the federal task force that would reduce the amount of water needed by the city.
(1) ______________________ (2) ______________________

26. Explain why water from Mono Basin is cheaper than water available from the Metropolitan Water District. ______________________
____

27. Water flowing from Mono Basin supplies Los Angeles with both water and ______________________.

Will the Lake Survive? [340]

28. What changing condition may cause Mono Lake to become the Dead Sea of California? ______________________

29. Assume that you are a federal court judge who has been assigned the case of Mono Lake vs. the City of Los Angeles. Your job is to decide if the Los Angeles Department of Water and Power has a right to divert water from Mono Lake if it will destroy the lake. Give your decision and write one or two paragraphs which supports your decision.

____
____
____
____
____
____
____
____
____
____
____
____
____
____
____
____
____
____
____
____

Name ______________________________ Date __________ Class Period ____

Unit IV — 2.3 Can We Save the Chesapeake?

Introduction [342]

1. According to Charles Robb, to insure that we remain a strong society we must preserve our __________________ and __________________ resources.

2. The area where fresh water meets salt water is called an ________________.

3. The Chesapeake is the largest, most complex **(bay/river)** in North America.

4. The river that supplies the Chesapeake Bay with most of its fresh water is the **(Delaware/Chesapeake/Potomac/Susquehanna)**.

5. The land that drains into a river or a bay is referred to as the basin or __________.

6. Most of the Chesapeake **(Basin/Bay)** lies in the state of Pennsylvania.

The People's Bay [342]

7. "The Bay" is the site of many industries. It is an ideal location because the bay provides an easy (a less expensive) way to __________________ raw materials and products.

8. Give two reasons why many people come to the Bay region.
   (1) ______________________ (2) ________________________

9. Before people came to "The Bay," the land was mostly _____________ or _____________. Now it is mainly _____________________________.

Habitat for Wildlife [343]

10. Grasses growing in the wetlands along "The Bay" provide __________ and ____________ for wildlife.

11. Runoff from the land carried _____________________ needed by the grasses growing in the wetlands surrounding "The Bay."

12. Wetlands are important _______________ for fish. Wetlands produce **(less/more)** food per acre than can be produced on Iowa farmland. This means that wetlands are **(less/more)** productive than Iowa farmland.

13. Estuaries are important ______________ factories. Identify three kinds of fish that depend upon estuaries. (1) _________________,(2) ________________, (3) __________________

14. Identify three kinds of shellfish that are dependent upon estuaries.
    (1)________________, (2) ______________, (3) ______________________

15. For some species of fish, the **food chain** begins in the __________________. For others, **life** begins in the ___________________.

Name ____ Date __________ Class Period ____

Fish and Wildlife in Trouble [343]

16. The Chesapeake is the major source of __________________ on the Atlantic coast.

17. As the human population has increased in the area of "The Bay," the water quality has _______________.

18. The oyster population in "the Bay" is being threatened by: (1) __________________ (2) _______________________________ (3) _______________________________

19. The population of this species of fish has declined so much that it is now listed as "threatened." **(shad/striped bass)**

20. Give two reasons why the populations of fish that are dependent upon "The Bay" have declined. (1) _______________________ (2) _________________________

21. Why is the stripped bass a "threatened species" but species like the menhaden is not, even though both species breed in "The Bay." Explain why.
____
____

22. What is the most economically important fish in the bay? _____________________
The main use of this species is to make __________________ that is used in _____________________ and __________ food.

23. The major producers in "The Bay" that are the basis of many food chains are the _____________________. More than _____ percent of the plants have disappeared.

24. Explain loss of plants in "The Bay" affects fish, shellfish and other small aquatic organisms. ____
How does the loss of plants affect waterfowl? ____________________________

Identifying the Problems [344]

25. What three questions did Congress ask the EPA to answer?
(1) ____
(2) ____
(3) ____

26. How long did it take the EPA to find the answers? ______________

27. Levels of certain pollutants in "The Bay' are increasing. Why do some pollutants remain in "The Bay" while others are carried out to the ocean?
____
____

28. What two types of activities have caused increases in the level of nutrients in "The Bay"? (1) ___________________________ (2) ___________________________

29. The two major nutrients found in "The Bay" are ____________________ and _____________________.

30. Some nutrients enter "The Bay" from **point sources**. These include __________________________ and __________________.

Name ______________________________ **Date** ________ **Class Period** ____

31. Some nutrients enter "The Bay" from **nonpoint sources.** Nonpoint sources refers to ______________ from farmland, from developed land and abandoned __________.

32. Another important nonpoint source of pollutants in "The Bay" is ____________. It is a source of __________ oxides, _______________, and toxic chemicals including __________ and ___________.

33. Identify the major point source and nonpoint source of the nitrogen and phosphorus that enter "The Bay" from the state of Maryland. (1) _______________ (point source) and, (2) _______________ (nonpoint source).

34. The high level of nutrients increases the growth of __________ in "The Bay."

35. What two factors are responsible for reducing the amount of sunlight available to the submerged plants? (1) _______________ (2) _______________

36. The growth of algae eventually causes **(a decrease/an increase)** in the growth of submerged plants due to the lack of _______________.

37. The dead algae and submerged plants begin to ____________ causing the level of ____________ to become a limiting factor for fish.

38. In the 1930s the level of dissolved oxygen created a problem during only two months. Today low levels of dissolved oxygen create problems for some organisms during _____ months of the year.

Toxic Substances [345]

39. Chemicals that harm living organisms are said to be _______________. Such chemicals include certain metals, such as _______________ and __________, and _______________ compounds, such as _______________.

40. Identify two **point sources** of toxic chemicals that enter "The Bay."
(1) ______________________ (2) ______________________

41. Identify two types of toxic chemicals that may enter "The Bay" from storm drains.
(1) ______________________ (2) ______________________

42. Four quarts of oil could form an oil slick that would cover the surface of an _____ acre lake (one acre is approximately the size of a football field).

43. Identify three types of organisms that may be smothered by oil dumped in the water. (1) _________________ (2) _________________
(3) ______________________________.

44. List three ways in which toxic chemicals used by the average homeowner might enter "The Bay." (1) ____________________ (2) ________________
(3) ______________________

Name ______________________________ Date ________ Class Period ____

45. A survey of organisms in areas contaminated with toxic chemicals reveals the number of species is **(less than/greater than/the same as)** the number of species in other areas. The populations are **(larger than /smaller than/the same size as)** the populations in areas that are less polluted.

46. Some toxic chemicals become concentrated in organisms because they are stored in ________. The concentration of the chemical found in fish is **(lower/higher)** than the concentration in aquatic insects.

47. The increase in concentration of toxic substances at higher levels in the food chain is referred to as ____________________________.

48. _____________ in rain may be dissolving ______________________ from the soil.

49. Studies of tissue taken from fish living in areas with high levels of toxic chemical have shown **(some abnormal cells/cells that are normal).**

Working Together to Save the Bay [346]

50. The purpose of the **Chesapeake Bay Agreement** is to ____________________
____

51. **Pretreatment** will remove ________________________ from waste water before it enters "The Bay."

52. In addition to industrial discharges of waste water, it may be necessary to pretreat ____________ water.

53. **(Nitrates/Phosphates)** enter "The Bay" when it is leached from cropland and animal waste. **(Nitrates/Phosphates)** are attached to particles of soil and doesn't usually enter "The Bay" unless _____________ occurs.

54. The **Best Management Practices** of reduced tillage and contour farming will reduce _______________ lost from farm fields and improve water quality.

55. How will soil testing and timing application of nutrients help both the farmer and "The Bay"?
Farmer: ____
Bay: ____

56. A stream should be separated from human activity by a strip of ______________ that is _______________ feet wide.

57. To reduce the load of nutrients entering "The Bay," we must have better treatment of __________________. Storm water **(should/should not)** flow through the sewage treatment plant.

58. In order to determine whether significant changes are occurring in "The Bay," the EPA and the states must __________________ physical and chemical conditions.

59. Identify three populations which are being "restocked" or "replanted" in "The Bay."
(1) ___________________________, (2) ____________________________, and
(3) ______________________________

Name ________________________________ Date _________ Class Period ____

60. Since scientists can not predict the success of the efforts to improve "The Bay," do you think that the government is wasting millions of dollars or is it worth the effort? Explain. ________________

Lessons Learned [348]

61. Explain what is meant by each of the following statements.

A. "The solution to pollution is **not** dilution.." ____________________________

B. "There is no away." ____________________________

C. "Political boundaries do not stop pollution." ____________________________

You Can Make a Difference [348]

62. What should you do with used oil or antifreeze? ____________________________

63. If you need to dispose of a toxic substance and you do not know the safe disposal method, what agency in your county will be able to provide the information you need ____________________________

Vocabulary Review

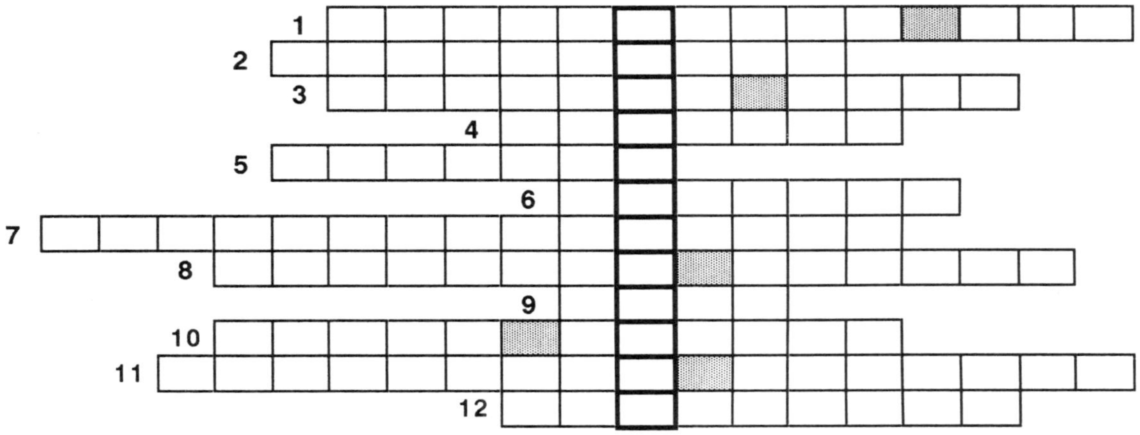

1 the largest estuary in North America.
2 measuring and recording physical and chemical conditions of an ecosystem on a regular basis.
3 species of fish in the Chesapeake Bay that is listed as "threatened."
4 area near the mouth of a river where fresh water and salt water mix.
5 a type of shellfish that depends upon estuaries.
6 _________ oil instead of dumping it down the drain.
7 the concentration of a chemical increases as it passes through the food chain.

Name ______________________________ Date ________ Class Period ____

8 runoff from farm fields and lawns is a ________ _________ of pollution.
9 abbreviation for those procedures used to help reduce agricultural pollution.
10 If you have a question concerning the disposal of a toxic substance, call the Cooperative ___________ ____________ in your county.
11 Wastewater from industries often contains _________ __________ like cadmium, lead, and zinc.
12 Phosphorus and nitrogen are plant ___________ that cause algae blooms.

Name ______________________________ Date __________ Class Period ____

Unit IV — 2.4 "Waters of the United States"

Introduction [351]

1. What agency of the government is responsible for enforcing the Clean Water Act? ______________________________

2. You must obtain a Section 404 permit in order to ______________________________ ______________________________

3. What government agency is responsible for issuing the permit? ______________ ______________________________

4. **"Waters of the United States"** refers to any waterway that is ____________. It also includes ______________ and __________ that flow into navigable waterways and ________________ beside any of these bodies of water.

5. What are the three factors that must be studied in order to identify land as **wetland**? (1) ______________, (2) __________________, (3) ____________________. Plants growing in a wetland are adapted for living in soil that is ______________ with water.

6. Identify two types of wetlands that are dry for much of the year. (1) ____________ ____________________ (2) ______________________________

   Identify the type of wetland that has water-logged soil but is seldom flooded. ______________________________

Water or Land? [351]

7. By storing flood water and slowing the flow of runoff, wetlands prevent ____________________ and _________________.

8. Wetlands decrease the ________________ of the shoreline during storms.

9. Water quality is improved because wetlands ____________________ the water.

10. The purpose of the **Swamp Lands Acts** passed in the 1800s was to __________ wetlands. The former wetlands were used for ______________, ______________, or ______________. More than _____ percent of the wetlands that existed in the 1800s have been drained and filled.

11. Wetlands provide _____________ for wildlife and plants, recharge the _________ ___________, provide _____________ and ___________________.

12. _______________________ of the original wetlands in the lower 48 states have been reclaimed. Most of the wetland loss is due to ___________________.

13. The loss of wetlands has **(slowed/stopped)**.

Riverside Bayview Homes, Inc. [352]

14. Why did the Corps of Engineers file suit against Riverside Bayview Homes, Inc.?

Name ______________________________ Date ________ Class Period ____

____

15. According to the final decision made by the U.S. Supreme Court:
Are wetlands a part of the "waters of the United States"? ________
Was the land owned by Riverside Bayview Homes, Inc. wetland? ________
Were the developers required to obtain a 404 permit? ________

Moving a Marsh [353]

16. The purpose of Section 404 of the Clean Water Act is to:
A. prevent development of the wetlands.
B. protect water quality and aquatic ecosystems.

17. **(True/False)** Most requests for Section 404 permits are denied.
(True/False) If there is no alternative, development of wetlands is allowed.

18. The Wisconsin Department of Transportation **(was/was not)** allowed to expand the four-lane highway across the marsh.

19. How did the Wisconsin DOT build the road without reducing the amount of wetlands? ______________________________
____

Swamp or Shopping Center [354]

20. The government agency that vetoed the Section 404 permit issued to the Pyramid Companies of Boston was the ______________________________

21. What reason was given for vetoing the permit? ____________________

Put it Back the Way it Was [354]

22. If a landowner or developer breaks the law by filling in wetlands without a Section 404 permit, they will most likely be:
A. fined and given a jail sentence. B. required to restore the wetland.

23. Steps that are required to restore a wetland include:
(1) ______________________________
(2) ______________________________
(3) ______________________________

Can Humans Imitate Mother Nature [355]

24. Why was a part of San Diego's Sweetwater Marsh destroyed? ____________
____
Did the 404 permit require the creation of a new marsh? ________

25. Five years after the new marsh was created, did it function like the old one? ______
What makes it difficult to create a new wetland that will function like the original wetland? ______________________________
____

26. According to the National Research Council, restoring wetlands is a "__________ and __________" process.

Name ________________________________ Date ________ Class Period ____

Unit IV — 2.5 Dam the Rivers or Let Them Run Free

The Army's Dam Builders [357]

1. The nation's waterways were made navigable by the ______________________

2. Today the Corps is responsible for maintaining the waterways and protecting the ________________.

3. In the past, wars have been fought over rivers, but today the battles over the use of a river are fought in ______________________.

4. Critics claim that the Army Corps of Engineers *under* estimates the cost of the ___________ and the value of the __________.

5. Critics claim the Corps *over* estimates the _____________ and the __________ of a dam.

6. The total cost of a dam is equal to the cost (in dollars) of building the dam plus the value of the _________________ that is lost. If the total cost is greater than the benefits gained from the dam, the dam should not be built.

Dams Provide Flood Control [358]

7. The area that would have been protected by the dams on the Meramec River was a **(densely/sparsely)** populated region near the city of __________________. The areas that would have been flooded by the dams were **(densely/sparsely)** populated.

8. The dams that had been proposed for the Meramec River were never built. Why not? ____

9. What "event" caused the COE to propose 14 dams on tributaries of the Brodhead Creek in Monroe County, Pennsylvania? ____________________________ How many of the dams were built? ________

10. The Tocks Island Dam proposed for the Delaware River was never built. If built, would it provide flood protection for Monroe County? ______ What was the purpose of the dam? ____

Save the River [358]

11. What recreational activities would be possible at Meramec Park Lake that are not possible on a free-flowing stream? ___________________________________ ____

    According to the COE did the area need additional flat-water recreational activities? ______ According to Missouri's Statewide Outdoor Recreation Plan, did the area need additional flat-water recreational activities? ______

12. According to the COE, would the Meramec Dam have improved the conditions for the canoeist? ______ According to opponents of the dam, would the dam have improved the conditions for the canoeist? ______

Name ______________________________ Date ________ Class Period ____

13. List the four purposes of the Tocks Island Dam and Lake Project.
(1) ______________________ (2) ______________________
(3) ______________________ (4) ______________________

14. The total cost of the Tocks Island Dam is equal to the cost (in dollars) of the dam and the value of the ______________________ that would be lost.

15. There were several reasons why the Tocks Island Dam has not been built. Two of these reasons are related to the weather. Explain.
____
____

16. The need for additional power **(was/was not)** demonstrated.
The need for free-flowing streams for recreation **(was/was not)** demonstrated.

17. Budget is always a consideration. The federal government did not have extra money for building dams because of the cost of the ______________________.

18. How did Congress protect the region of the river that would have been covered by the Tocks Island Lake? ______________________
____

Dams Provide a Water Supply [359]

19. **(Lakes/Wells)** are the source of water for people near the Meramec River.

20. According to the Corps of Engineers, what advantage does a lake offer over wells?
____

21. List the four states that share the water resources of the Delaware River Basin.
(1) ______________________ (2) ______________________
(3) ______________________ (4) ______________________

22. Before 1960, disputes over water in the Delaware River Basin were settled by the ______________ Today decisions that affect water quality and water use are made by the ______________________.

23. The job of Delaware River Basin Commission is to manage ______________ ______________ in the Delaware ______________. The commission must:
(1) set policy for the use of the ______________
(2) set standards for ______________
(3) approve projects that affect ______________ in the watershed including the building of dams.

The DRBC **(has/has not)** approved dams for the purpose of ______________
____

24. New York City **(is/is not)** in the Delaware River Basin. Nearly ______ percent of the water used by the residents of New York City is supplied by the Delaware River Basin.

25. The ______________ River is the only major source of surface water in the Southwest U.S.

Name ________________________________ **Date** __________ **Class Period** ____

26. Identify three major U.S. cities that depend on water from the Colorado River.
(1) ____________________ (2) ____________ (3) ____________
Do any of these cities lie in the Colorado Basin? _______ If so, which one(s)?
____

27. Why does the Columbia River Basin receive so little rainfall each year?
____
____

28. __________________ dam has changed the part of the Columbia River that lies in the U. S. from a free-flowing river into a lake.

29. Much of the water stored behind the dam is used for ______________________.

30. Most of the water in the Columbia River fell as snow on **(Canada's Rocky Mountains/Washington's Cascade Mountains)**.

Dams Provide Water to Increase the Natural Flow [361]

31. During a drought, people living in Camden, New Jersey, are concerned that their water will be contaminated with __________.

32. The ________________ describes the place where fresh water from a river meets salt water from the ocean.

33. The salt front on the Delaware River is usually _____ miles downstream from Camden, New Jersey, near the city of ____________________, Delaware.

34. Water treatment plants are not normally designed to remove salt because the **(process is expensive/technology is not available)**.

35. The source of drinking water in Camden, New Jersey, is **(wells/the river)**.

36. Explain how Camden's water supply might become contaminated with salt.
____
____

37. Are industries that do not drink the water or use it in food products concerned about the salt content of the water? _____ Why or why not? ______________________
____

38. Does New York City's use of water from the Delaware River Basin increase the chance that the water supply for Camden and Philadelphia will be contaminated with salt? _____ Explain. ____
____

39. In addition to the reservoirs built by New York City, reservoirs on tributaries of the Delaware River store water for release during low-flow periods. According to the Delaware River Basin Commission, is the storage capacity adequate to insure protection of water supplies for cities like Camden and Philadelphia? _____

Name ______________________ Date __________ Class Period ____

40. List three factors that might cause the Delaware River Basin Commission to decide that additional dams should be built on the Delaware River.
(1) ______________________ (2) ______________________
(3) ______________________

Dams for Navigation [362]

41. Why have rivers been such an important means of transportation?
____
What city claims to be the largest inland port? ______________________
What river begins at this port? ______________________

42. What made navigation of the Ohio River near Louisville, Kentucky, so dangerous?
____

43. How did the Army Corps of Engineers improve navigation through this section of the river? ______________________
____

44. Explain how a **lock** allows a barge to safely move through an area where a waterfall or a dam would otherwise prevent travel.
____
____

Dams for Hydroelectric Power [362]

45. New fossil-fueled power plants are 40 percent efficient. The efficiency of a new hydroelectric power plant is _____ percent.

46. What region of the United States produces most of its electricity at hydroelectric power stations? ______________________

47. In addition to the production of electricity, a large amount of water in the Columbia River is used for ______________________

48. What are the two limiting factors for further development in the Pacific Northwest?
(1) ______________________ (2) ______________________

49. In addition to the cost (in dollars) of additional dams built along the Snake River, the environmental cost would be loss of ______________________
____

50. If you were a U.S. senator, would you vote for a bill which would allow additional dams to be built on the Snake River in the Hells Canyon National Recreation Area?
_____ Why or why not? ______________________
____

51. Give the position of each of the following groups of people toward construction of dams and briefly explain the reason that supports the position.
Salmon fishermen: **(against/for)** ______________________
Aluminum manufacturers: **(against/for)** ______________________
Cree Indians: **(against/for)** ______________________
____

Name ______________________________ **Date** ________ **Class Period** ____

Dam Safety [364]

52. What events caused the President of the United States to call for an inspection of the nation's dams? ______________________________

53. What agency of the federal government is responsible for the inspection of dams? ______________________________ How many dams must be regularly inspected? __________

54. What two factors must be considered in order to determine if a dam is safe?
(1) ______________________ (2) ______________________

55. Identify two geological features that make an area unsuitable for dams.
(1) ______________________ (2) ______________________

56. A geologist is asked to select the best site for a dam. One of the possible sites is in an area where the bedrock is granite. The rock structure at the second site being considered is limestone. Which site would the geologist probably recommend?
____

Name ______________________________ Date __________ Class Period ____

Unit IV — 2.6 Rivers on the Land

Introduction [366]

1. Quickly review the water cycle by identifying the processes described below. If necessary refer to Unit I, Section 4.1.
   ______________ Rain disappears into the soil.
   ______________ The water journeys across the land to a nearby stream.

2. The **load** carried by water refers to the ________ and _______ particles that are eroded from the land.

The Water Drops its Load [366]

3. The low, flat land that lies along a river is called the ______________. Floods help to restore the supply of ________________.

4. Floods usually **(cause erosion of/deposit soil on)** floodplains.

5. Land along the banks of rivers is often **(good/marginal)** farmland because periodic floods create a thick layer of __________ topsoil.

6. Both on flood plains and at the mouths of rivers, the slope of the land is usually **(gentle/steep)**.

7. New land created at the mouth of a river is called a __________. In the past, soil eroded from states in the Mississippi watershed has been deposited at the river's mouth increasing the size of the state of ______________. But now the delta is ____________ due to flood-control structures built to protect ____________.

Cities Along the River [367]

8. Early settlers chose sites beside rivers for their villages because the river provided.
   (1) ______________________ (2) ______________________
   (3) ______________________

9. Industries chose sites along rivers because the rivers provided both
   (1) ______________________ (2) ______________________

10. Most power plants along the Ohio River use **(coal/water)** to supply energy needed to make electricity.

11. **(True/False)** Coal-fired power plants use large amounts of water.

12. In addition to water, a site chosen for industry must also have a good supply of ____________________.

Raindrops Falling on the City [367]

13. As development increases the infiltration of water __________________.

14. As the natural flow of water is disrupted by development, the runoff is collected and carried in ____________________.

Name ______________________________ Date ________ Class Period ____

15. The low-lying areas of cities will flood **(faster/slower)** than low-lying areas in undeveloped areas. This occurs because of the reduced amount of ____________.

Straight and Narrow Channels [367]

16. Describe the following characteristics of human-made channels as compared to natural channels of rivers and streams.
Length: Human-made channels are ________________ than natural channels.
Path: Human-made channels are ________________ than natural channels.
Slope: Human-made channels are ________________ than natural channels.

17. The volume of water that can be carried by a human-made channel is **(greater than/less than)** the volume carried by the natural channel it replaced.

18. The amount of flooding in the area where the human-made channels are located is **(increased/reduced)**. Downstream from the channelized area the possibility of flooding is **(increased/reduced)**.

Johnstown, PA — City on a Floodplain [368]

19. The city of Johnstown has the distinction of being the city where floods have **(occurred more often/caused greater loss of life)** than in other cities in Pennsylvania.

20. The massive destruction and loss of life in flood of 1889 was due to **(failure of a dam/a record rainfall)**.

21. The repairs made on the dam **(were inspected and certified as safe/did not meet engineering standards)**.

22. The destruction and loss of life in the flood of 1936 was due to **(failure of a dam/record rainfall)**.

23. To reduce the chance of flooding, the U.S. Army Corps of Engineers channelized or ______________ the mountain stream and built ________ to protect the city.

24. After the changes made by the U.S. Army Corps of Engineers in 1936, Johnstown was flood-free for more than ______ years.

25. The 1977 flood was caused by **(failure of a dam that did not meet safety standards/a record rainfall)**.

26. The changes made by the U.S. Army Corps of Engineers in 1936 **(prevented/reduced the amount of)** flooding of Johnstown in 1977.

27. Should dams be built according to safety standards that will prevent the dam from collapsing during a "once in five hundred year" storm? ________ If not, how should the people downstream be protected? ____________________________
____

28. How could the U.S. Army Corps of Engineers claim that their flood protection project was a success when more than 75 people died? ____________________
____

Name ______________________________ Date ________ Class Period ____

The Value of a Floodplain [369]

29. List the two natural functions of a floodplain.
(1) ______________________________
(2) ______________________________

30. If left alone the floodplain serves two important functions that help humans. Why do the owners of floodplains often want to see this land developed? ____________
____

The Cost of Using a Floodplain [370]

31. The amount of property damage in the Johnstown area caused by the flood of 1936 was **(greater/less)** than the flood of 1889.

32. After the COE flood protection project, the amount of damage caused by the flood in 1977 was **(greater/less)** than the flood of 1936.

33. The major cause of disasters is **(earthquakes/floods)**.

34. An area that is declared a disaster area by the President of the United States, receives **(help from the Army Corps of Engineers to rebuild/financial assistance from the federal government)**.

35. Identify major cities in your state which are built on floodplains.
____

36. Identify three ways that cities were protected from flooding by the national flood protection program. (1) ______________ (2) ______________
______________ (3) ______________

37. **(True/False)** Thousand of homes and businesses in areas protected by levees have been flooded.

38. List three ways that the life of people in flooded areas is disrupted even though their home is not flooded and no lives are lost.
(1) ______________ (2) ______________
(3) ______________

Different Approaches to Flood Control [371]

39. While a farmer loses a crop because of flooding, the flood also gives the farmer something to compensate for the loss. What does the flood provide?
____

40. Give three reasons why the city council may feel it is necessary to develop the floodplain. (1) ______________ (2) ______________
(3) ______________

41. List the three uses of the floodplain that were approved under the land-use plan adopted by St. Louis County in 1960. (1) ______________
(2) ______________ (3) ______________

Name ______________________________ Date ________ Class Period ____

42. If you had been a voting member of the St. Louis County government in 1970, would you have voted to rezone the 1200-acre site on the Missouri River floodplain to allow the building of Earth City? ________ Why or why not? ____________
____
____

43. How do you think that the construction of the industrial complex at Earth City affected the communities down stream during the "Great Flood of '93"?
____
____
____

44. The Corps of Engineers recommended that to protect themselves from flooding Boston and Cambridge should **(build more reservoirs/ buy wetlands/ increase the height of the levees)**. The plan was called the ____________ ________________________.

An Act of Congress [372]

45. Describe how the presence of flood control structures may contribute to the increased loss of life and property? ______________________
____

46. Give two ways the Federal Flood Disaster Protection Act will decrease the amount of federal money spent for damages due to flooding.
(1) ____
(2) ____

47. What advice do the experts offer today regarding the best approach to control?
____
____

48. Explain the statement: "The river and the land are partners."
____
____
____

Name ________________________________ Date __________ Class Period ____

Unit IV — 2.7 The Mississippi — River on the Move

Introduction [375]

1. The Mississippi River begins in ______________________________ and flows to the **(Atlantic/Gulf of Mexico/Pacific)**.

2. The Mississippi River Basin includes watersheds in ______ states and two of __________________ provinces.

The River as a Highway [375]

3. List three ways that the Army Corps of Engineers changed the Mississippi River.
(1) ____________________________ (2) ____________________________
(3) ____________________________

4. How have cities and industries changed the river? ______________________

5. List the two responsibilities that the federal government has assigned to the Army Corps of Engineers.
(1) ____
(2) ____

Changing Rivers [375]

6. Explain how the steamboat Sultana, that sank in the Mississippi, became buried beneath a field in Arkansas. ______________________________
____

7. Describe the path of old rivers that flow across land with a gentle slope.
____

8. Part of the water carried by the Mississippi River flows into this Louisana River ____________________________.

The River versus the Corps [376]

9. The purpose of the river control structure built by the Corps is to:
___ prevent water from flowing into the Atchafalaya River.
___ prevent water from flowing into the Mississippi River.

10. If the Mississippi River is allowed to change its path it will affect many people. Describe how it will affect people in the following areas:
the Midwest: ____________________________________
New Orleans: ____________________________________
Morgan City, LA: ________________________________
Pennsylvanis, New Jersey, New York: ________________
____

11. The cost of maintaining the flow of water in the Mississippi is $5 million/year. If you could vote on each item in the federal budget, would you vote for or against this item? _________ Justify your vote. ______________________
____

Name ______________________________ Date _________ Class Period ____

12. Do you think that the Corps will be able to maintain the flow in the Mississippi or will the water eventually take the easier path down the Atchafalaya River? _______ Explain. ____
____

Name ______________________________ Date _________ Class Period ____

Unit IV — 2.8 Cleaning Up the Rivers

Introduction [377]

1. Identify two major events that increased the waste entering the nation's rivers.
(1) ____________________________ (2) ____________________________

Reviving the Thames [377]

2. In the mid-1800s many people living in London died from ____________ when they ate food or drank water that was contaminated with human waste.

3. In the 1800s the Thames River was an open sewer. To improve the quality of water in the river, sewer pipes were installed to carry the waste to _________________. Today sewer pipes carry the waste to __________________________.

4. Today **(few/many)** species of fish are found in the Thames. The estuary at the mouth of the Thames River is one of the world's **(healthiest/most polluted)** estuaries.

5. Evidence of the healthy environment in the Thames River is the return of the _________, a fish that requires clean water with a high level of dissolved oxygen.

Restoring America's Rivers [378]

6. The first sewers installed in New York City carried sewage to the ___________.

7. Would salmon have survived in the Hudson River in the early 1970s? _________

8. Pollution in the Charles River caused the disappearance of many species of fish including the ___________.

9. Each spring shad migrate up the Delaware River to spawn. Is it likely that anyone fishing in the spring of 1952 caught a shad? ______ Explain. ________________
____

10. What was the first step taken to improve the quality of water in Oregon's Willamette River? ____

11. What industry produced much of the oxygen-demanding waste in the Willamette River? ______________________________________

12. How does the level of dissolved oxygen in the Willamette River today compare to the DO level recorded in the 1960s? __________________________

13. What major industry in Maine polluted the state's rivers? _______________

14. Why do some people object when laws are passed that will produce a cleaner environment? ____

15. The quality of water in Maine's rivers is indicated by the return of _____________ ___________.

Name ______________________________ **Date** ________ **Class Period** ____

16. What industry has benefited from the cleaner rivers? ______________

Fishable and Swimmable Rivers [379]

17. The goal of the 1972 Clean Water Act was to have streams and rivers in which both ________ and ____________ could safely swim.

18. To meet the water quality standards, the law requires:
cities to: treat ____________
industries to: use the ____________________ technology to clean waste water.

19. Do all rivers and streams meet the water quality standards set in 1972? _____
Is the water in Boston Harbor "fishable and swimmable"? _____

At the Sewage Treatment Plant [380]

20. As sewage enters the sewage treatment plant, it passes through a ____________ or ____________ before it enters a large tank.

21. Sewage is held in the first tank to allow the **(chemicals to be properly mixed/solids to settle out)**.

22. In the settling tanks, solids that sink to the bottom are called ____________. A skimmer removes the scum and ____________ that float on the surface. The remaining waste water is called the ______________.

23. Primary treatment of sewage is a **(biological/physical)** process.

24. Municipalities in the U.S. are required to provide **(both primary and secondary/only primary)** treatment.

25. The **activated sludge process** is **(less/more)** expensive than the trickling filter process, but it requires **(less/more)** space than the trickling filter process.

26. In the **activated sludge process** _______ is pumped into the tanks with sewage.

27. ____________ need oxygen for breaking down or digesting ____________ in the sewage.

28. After several hours in the aeration tanks, the waste water is pumped to secondary ____________________. Some of the **(effluent/sludge)** is recycled to the aeration tanks so that the millions of ___________ it contains can digest more organic matter.

29. In a **trickling filter** the waste water is sprayed over a bed of **(sand/stones)** that are covered with ________ and ______________.

30. The purpose of primary and secondary treatment is to remove ______________ from the water. Primary and secondary treatment can remove _____ percent of this matter.

Name ______________________________ Date _________ Class Period ____

31. The **(effluent/sludge)** from the primary and secondary settling tanks is pumped into heated tanks called __________. More organic matter is removed by __________ that **(do/do not)** use oxygen.

32. In the digesters, the bacteria produce the waste product __________. It is burned to provide __________ for the digesters and sometimes __________ nearby.

Sludge — It's Not a Dirty Word [381]

33. Describe how each of the following cities disposes of the digested sludge.
New York City (prior to 1992): ______________________________
Chicago: ______________________________
Hollands Landing, Canada: ______________________________
Lufkin, Texas: ______________________________
New York City (constructing facilities to): ______________________________

34. Where space is available, sludge may be spread on beds of __________ to dry. In cold climates the drying beds are located in __________. Once the sludge is dry it can be used as fertilizer or it can be __________.

35. Explain why it is unlikely that the digested sludge will cause disease. __________
____

36. If it were available, should you use sludge on your garden?_______
Should you buy vegetables from a farmer that used sludge on these crops?_______
Is it safe to buy pork from a farmer who used sludge as fertilizer to grow the corn fed to the pigs? _______

37. To calculate the correct amount of sludge to use, the farmer must know the amount of nutrients in the __________ and the __________. If the farmer applies too much sludge to the land, __________ may pollute the water supply or streams located nearby.

38. What organism is being used in several countries to compost sewage sludge? __________ In addition to sewage sludge, identify two types of waste which may be composed by this process.
(1) ______________________, (2) ______________________

The Need for Pretreatment [383]

39. Sewage sludge and soil must be tested for __________ that might affect plant growth.

40. The heavy metal __________ might cause problems for humans and other animals at levels that are too low to affect plant growth.

41. __________ is the source of most heavy metals found in sludge.

42. Heavy metals may **(be removed by/interfere with)** primary and secondary treatment processes at the sewage treatment plant.

43. Bacteria do not have the proper __________ needed to break down toxic wastes.

Name ______________________________ **Date** __________ **Class Period** ____

44. Even though some heavy metals may not interfere with treatment processes, they may become concentrated in the sludge or enter aquatic ecosystems with the effluent. To prevent problems caused by heavy metals, industries must __________ waste water.

45. Identify the process which is responsible for each of the following:

    Pre = pretreatment **P/S = primary and secondary sewage**

    _____ A. removal of organic matter.
    _____ B. remove heavy metals.
    _____ C. decrease biological oxygen demand (BOD).

Chlorination [383]

46. Before it is discharged into a lake or stream, ________________ is added to kill any bacteria in the effluent.

47. **Trihalomethanes (THMs)** are formed when chlorine combines with __________ ____________ in the water. THMs cause **(birth defects/cancer)**.

48. Since chlorine is harmful to organisms in the environment, sewage treatment plants in some regions are allowed to release effluent without chlorination during the ______________ months.

49. Sewage treatment plants that release effluent near ____________________ must always chlorinate effluent.

Combined Sewers [384]

50. **Combined sewers** carry both _________________ and __________________.

51. During storms in cities with combined sewers, storm water and _____________ ______________ flow into the river.

52. Are the sewers in your town or city combined or separate (storm and sanitary)? ______________________

The Effluent — Additional Treatment [384]

53. Even after primary and secondary treatment, the effluent leaving some sewage treatment plants does not meet EPA standards for _________________ and _________________.

54. If effluent with high levels of these nutrients enters streams it may cause ________ ____________.

55. To avoid pollution of streams, effluent with high levels of nutrients might be used for ____

56. Effluent might be diverted into a natural or artificial ________________ where plants would remove the excess nutrients before infiltration occurs.

Name ______________________ Date __________ Class Period ____

Lagoons and Aquaculture [384]

57. Indicate whether each feature listed below is an **advantage A** or **disadvantage D** of using a lagoon as a means of treating sewage.
_____ space needed _____ expense _____ maintenance

58. Explain how a lagoon is like a primary treatment tank at a sewage treatment plant.
____

59. Describe how a lagoon is like a trickling filter. ______________________________
____

60. Describe how each of the following problems associated with lagoons might be solved.
Odors: ______________________________
Mosquitoes: ______________________________
Water leaving lagoon still contains nutrients: ______________________________

61. Identify plants which help clean waste water in lagoons in:
Warm climates: ______________________________
Cold climates: ______________________________

62. Explain how a farmer who raises beef cattle might use a lagoon to solve pollution problems and decrease costs of raising the beef. ______________________________
____

63. At Muskegon, Michigan, the wastewater is stored in lagoons and then used to ______________________.

On-site Sewage Treatment [385]

64. The most common type of waste water treatment in rural areas is the __________ – _________ absorption system.

65. Sewage and waste water enter the ___________ tank where _____________ break down organic matter. Then the effluent flows into a ________________ system or drain field.

66. Describe how the process that occurs in a septic differs from:
Activated sludge tanks: ______________________________
Digesters: ______________________________

67. The size of the septic tank is determined by the number of **(children/bedrooms)** in the home. If the tank is the proper size the sewage will remain in the tank for a minimum of ______ hours.

68. The size of the soil absorption field is determined by the:
(1) ___________ and the (2) ______________________ rate

69. The **percolation rate** refers to the time required for water to move through an inch of soil. The percolation rate must be more than _____ but less than ______ minutes per inch.

Name ______________________________ Date ________ Class Period ____

70. For a subsurface soil absorption system, the soil must be the correct type and the correct __________.

71. If the natural soil does not meet the minimum depth requirement, the absorption system may be placed in a layer of __________ above the surface of the soil.

72. The perforated drain pipes are always placed in a layer of __________ and covered by ______________________ before the entire drain field is covered with a layer of soil.

73. To prevent clogging of the drain field, __________ should never be poured down the drain.

Inside the Septic Tank [387]

74. About _____ percent of the solids that enter the tank are digested by the __________. The undigested solids become __________.

75. Explain why large amounts of disinfectants should not be poured down the drain.
____

76. When discussing the plans for their new home, the owners insist on a garbage disposal. How will this affect the septic system that the builder is planning to install? ____

77. The undigested solids must be pumped out of the septic tank. If it is not emptied regularly, the solids may ______________________________.
The mixture of liquid and solids removed from a septic tank is called __________
How often the tank must be pumped depends upon (1) ________________
and (2) ______________________________.

78. Is **septage** as safe as sludge from a municipal sewage treatment plant? _______.
Why or why not? ______________________________

79. The disposal site for septage is usually ______________. The texture of the soil at the disposal site should be some type of ______________.

80. For each of the following features, describe a site that is suitable for septage disposal if possible. If the site is not suitable write NO.
Bedrock: ______________________________
Water table: ______________________________
Slope: ______________________________
Sinkholes: ______________________________
Wetlands: ______________________________
Well or Spring: ______________________________
Temperature: ______________________________

81. Should a farmer allow beef cattle to graze in a pasture where septage has been sprayed? ______________________________
Should a farmer allow dairy cattle to graze in a pasture where septage? ______
Is it acceptable for a farmer to grow tomatoes in a field where septage has been sprayed? ______

Name ______________________ Date ________ Class Period ____

82. Why is the use of lagoons for disposal of septage being phased out in areas with limited farmland? ______________________
____

83. In Harwich, MA, septage is treated in a solar aquatic greenhouse where organic matter is removed by ____________ and ____________ ecosystems. The bacteria in the effluent are killed with ____________.

Is it Worth the Cost? [388]

84. Is it possible to have rivers which are "fishable and swimmable"? ______
If so, is it worth the cost? ________

85. Why was ocean dumping of sludge banned? ______________________
Does the increased cost of land disposal justified? ______ Explain.
____
____

Vocabulary Review

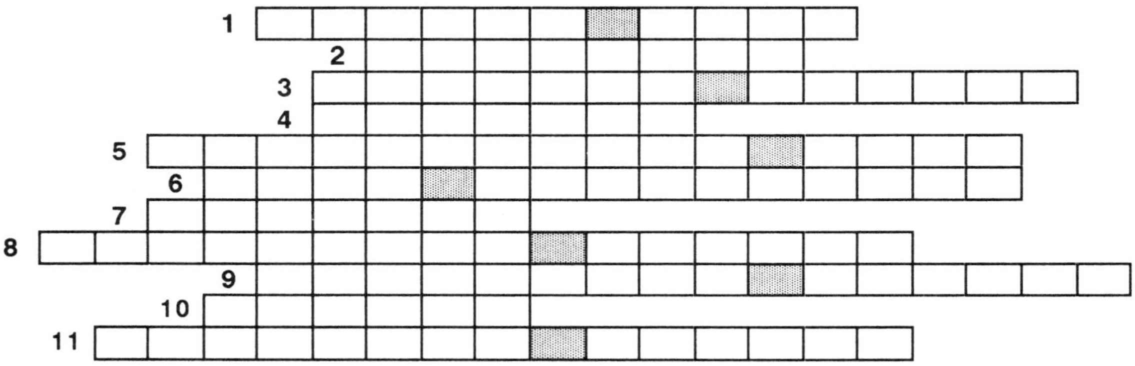

1 holds waste water before it enters the soil absorption field.
2 waste water that remains after solids are removed by settling at a sewage treatment plant.
3 the component of sewage that is removed by primary and secondary treatment.
4 before entering the sewage treatment plant, sewage passes through a grinder or __.
5 the speed that water moves through the soil.
6 the drain field for an on-site sewage treatment system.
7 gaseous waste product produced by bacteria during the digestion of organic matter.
8 secondary sewage treatment process that requires air pumped into the tank.
9 a less expensive means of sewage treatment occurs when waste water is held in ponds or ___ for 20 - 30 days.
10 this type of sewer system will not function properly during storms because of the volume of waste water that must be treated.

Name ______________________ Date __________ Class Period ____

Unit IV — 3.1 Poisons in the Water

Introduction [392]

1. List three ways in which synthetic chemicals have made our lives better.
   (1) ______________ (2) ______________
   (3) ______________

2. What problem is created by the manufacture of products?
   ______________________________.
   ______________________________.

3. List three sites that have been used to dispose of toxic chemicals (hazardous wastes). (1) __________ (2) __________ (3) __________

4. What natural resource is threatened by landfills and lagoons that are leaking?
   ____

5. Congress responded to the situation at Love Canal by creating a Superfund for ______________ and set strict regulations for ______________________________.

The Lesson Learned at Love Canal [393]

6. Once Love Canal was no longer needed as a canal, it was used for __________
   ______________________________.

7. After the canal was filled, a __________ and __________ were built on the site.

8. What health problems were reported? ______________
   ______________________________.

9. How did President Carter respond? ______________
   ______________________________.

10. What happened to the community? ______________
    ______________________________.

11. To insure that the chemicals are not leached from the site, the dump has been covered with ________ and water draining from the site goes to a special ______________..

12. What was the actual risk to the residents? ______________

13. Would you buy one of the homes from the government now? ________

Woburn's Wells [394]

14. When people living and working in the same area develop the same disease, the occurrence is called a **disease** ______________.

15. What disease cluster was found in Woburn, Massachusetts? ______________

Name ______________________________ Date ________ Class Period ____

16. Lawsuits filed by the families claimed that the occurrence of the leukemia cluster was due to the presence of ____________ chemicals in the town's _________.

17. The court found that:
( ) The company was responsible for pollution of the wells.
( ) The deaths were caused by chemicals polluting the wells.

18. A study made by Harvard's School of Public Health showed:
( ) A significant increase in leukemia.
( ) No significant increase in leukemia.

19. The Harvard study also found a link between the water and:
( ) The health of elderly people who drank the water.
( ) The health of infants born to women who drank the water.

20. Industrial chemicals threaten ______________ of American cities with populations over 10,000.

21. Which of the following statements is true?
( ) More studies are needed on long-term effects of low levels of pollutants.
( ) Studies have shown that low levels of pollutants are not dangerous.
( ) Studies have already shown that low levels of pollutants are dangerous and what we need to do now is to take action.

What the Fish Tell Us [394]

22. The major source of most of the PCBs in Lake Superior is polluted ________. Contamination of water by chemicals that are washed from the atmosphere by rain or snow is called ______________________________.

23. Large lake trout in Lake Superior are contaminated with _________. Some lakes in Florida, Wisconsin, Michigan and other states contain high levels of __________.

24. Based on the warning included with Michigan's state fishing licenses, what group of people will be most affected by eating fish that contain high levels of toxic chemicals? ________________________

25. The National Cancer Institute reports that:
( ) Carcinogens affect fish livers in the same way that they affect human livers.
( ) Areas with high numbers of fish tumors also have high numbers of human cancers.
( ) Both of these.

26. What kind of fish are being used by the EPA to identify cancer-causing chemicals?
____

27. Why are these fish ideal laboratory animals? ________________________
____.

28. EPA scientists hope to use these minnows to determine the level of specific chemicals that will harm ________________.

Name ______________________________ **Date** ________ **Class Period** ____

Water, Water, Everywhere, but... [395]

29. Why does New York City transport water from upstate when it is surrounded by water? ____

30. Water is transported from the upstate reservoirs to the city in pipes that are so large they are called ____________________.

New Orleans — City on a River [395]

31. What is the source of drinking water for New Orleans? ____________________

32. A 1974 study showed that New Orleans and several other communities using the same water supply had higher rates of: ___ leukemia ___ bladder cancer ___birth defects ___ gastrointestinal cancer

33. In an EPA study the THM found in highest concentration in New Orleans' drinking water was ____________________.

34. THMs (trihalomethanes) are formed when organic chemicals react with ___________. Scientific studies have linked THMs to _______________ and _______________ cancer in humans.

35. When expectant mothers are exposed to levels of THMs greater than 80 ______ the survival of the baby is in jeopardy. This concentration is **(higher/lower)** than the federal standard for drinking water.

36. To reduce the risk from THMs, New Orleans stopped using ______________ as a disinfectant and began using ________________ which is safer but more ________________. Some cities may use ____________________________ to remove the THMs.

Problems with Underground Storage Tanks [396]

37. The major industry in the Santa Clara Valley is the manufacture of ____________ ____

38. The major source of air pollution in Santa Clara County is the ________________, but the major threat to the valley is **(air/water)** polluted with ________________ ________________ used in the manufacture of high-tech equipment.

39. In many states the ground water supply is being threatened by ______________ which is leaking from underground storage tanks. The law required that all underground storage tanks be replaced by ______________________________ tanks by 1998.

40. While most of the gasoline can be pumped out, certain chemicals such as ________ dissolve in the water. This chemical causes **(birth defects/cancer)** in **(humans/laboratory animals)**.

Organic chemicals, including benzene, can be removed by using an activated ____________ filter.

Name ______________________ Date ________ Class Period ____

41. The disadvantage of using activated carbon filters is that they:
( ) do not remove all of the benzene
( ) are expensive

Down on the Farm [397]

42. The product of modern technology that threatens water beneath the Nebraska Sandhills is the ______________________.

43. The pollutant that threatens water supplies in the Nebraska Sandhills and other farming areas is ______________________.

44. Nitrogen is dangerous when it is in the form of **(nitrates/ nitrites)**.

45. The age groups affected by high levels of nitrates in the water is **(adults/children from birth to one year old/children prior to birth/older children/the elderly)**.

46. Nitrites affect humans and farm animals by:
( ) reducing the ability to fight infection.
( ) destroying the lining of the digestive system.
( ) decreasing the ability of the blood cells to carry oxygen.

47. Water with high levels of nitrate is not dangerous for adult humans and mature farm animals because the stomach contains high levels of _________ that kills the _______________ that change the nitrate into _______________.

48. Water with more than ______ mg/l nitrates should not be given to infants. If it has more than ______ mg/l nitrates it may be harmful for certain kinds of farm animals.

49. In some agricultural areas ground water supplies are also contaminated with ____________________.

50. Pesticides must be tested to determine how they affect living organisms, and for their ability to ________________ and contaminate ground water.

51. Check the true statement. Contamination of ground water supplies:
( ) will not occur if pesticides are applied according to directions given by the manufacturer.
( ) has been known to occur where pesticides were applied according to directions given by the manufacturer.

52. To prevent contamination of ground water supplies, the use of certain pesticides may be restricted to certain types of **(crops/soil)**.

Problems at Kesterson [398]

53. Land that is irrigated with ground water may become unsuitable for farming because of _________ deposited when the water evaporates. The buildup of salts and toxic elements in the soil from irrigation is ______________________.

54. In the San Joaquin Valley, a layer of ________ traps the irrigation water and a drainage system was installed.

Name ________________________________ Date __________ Class Period ____

55. Original plans called for a canal to carry waste-water to the ________________ ____________. The canal was not completed because of **(a lack of funds/ lawsuits filed by environmental groups)**.

56. The waste water was dumped into ____________________________ at the Kesterson National Wildlife Refuge.

57. Biologists became aware of problems when they noticed deformities in young **(birds/fish)**. Studies showed that the deformities were due to high levels of ________________ in the water. The mineral salt had leached from the ________.

58. Identify one problem associated with each of the following solutions to the problem at Kesterson.
Stop irrigation: ____
Finish the canal: ____

59. To minimize the problem in other locations, scientists developed new irrigation techniques that ____.

Lead [399]

60. Identify a health problem caused by high levels of lead in each of the following groups:
Children: ____
Women: ____
Men: ____

61. The major source of lead in drinking water is:
( ) air pollution
( ) chemicals used in agriculture
( ) corrosion of plumbing

62. The EPA drinking water standard for lead is 5 ________ in a sample taken at the **(tap/water treatment plant)**.

Under the City Streets [399]

63. It is ____________________ to properly dispose of hazardous wastes. Some companies, hired to dispose of hazardous wastes, ____________________ dump the waste in storm sewers or the ocean.

64. What is the goal of the Los Angeles special Strike Force team? ________________ ____

65. Judges are imposing heavy ____________ and sometimes ____________ terms on company officials who plan schemes to ______________ dispose of hazardous wastes.

Facing the Issue [400]

66. Twenty years ago our rivers were contaminated with ____________, today the water supply is threatened by the use of __________________________.

Name ______________________________ Date ________ Class Period ____

67. Toxic chemical enter rivers and lakes from point sources—plants that produce materials such as ______________________________ Nonpoint sources of toxic chemicals are ______________ from farms and cities and from ________ pollution.

68. List two ways that chemical companies can reduce the amount of hazardous wastes that might contaminate water supplies.
(1) ______________________________
(2) ______________________________

69. Is it possible to reduce pollution and save money? ________

70. Why did Minnesota limit the amount of cadmium, mercury and lead in packaging?
____
____
____

Name ______________________________ Date ________ Class Period ____

Unit IV — 3.2 Protecting the Groundwater

Introduction [403]

1. Which of the following best statements best reflects the opinion of the chemical industry:
   ( ) It is important that ground water be protected from toxic chemicals.
   ( ) Toxic chemicals in ground water are not a major problem because filters are available to remove them from water used for drinking.

2. Which of the following is less expensive?
   ( ) preventing pollution of ground water.
   ( ) cleaning polluted ground water.

Deep Well Injection —Is It the Answer? [403]

3. The chemical that caused the "rotten egg" odor in the state park near Erie, Pennsylvania, was ______________________.

4. The black liquid that had been stored in the Bass Island Formation was liquid waste from **(an oil/a paper)** company.

5. **Deep-well injection** involves pumping wastes into a layer of __________ rock located deep beneath the ground.

6. When injecting wastes into deep wells, the outer pipe is filled with __________. The purpose of this material is to prevent contamination of any ____________ the pipe passes through.

7. Oil companies use deep-well injection to dispose of ____________________ when drilling oil wells.

8. Identify three ways that toxic wastes injected into deep wells might reach the ground water.
   (1) ____
   (2) ____
   (3) ____

9. List two types of information that are needed to determine if a site is suitable for deep-well injection of toxic wastes.
   (1) ____
   (2) ____

The Problem with Landfills [404]

10. How does disposal at a **sanitary landfill** differ from the older "dump" sites where garbage was thrown on a pile and left to rot? ____________________
    ____

11. ________________ wastes should not be placed in a sanitary landfill because
    ____

Name ______________________________ Date __________ Class Period ____

12. Toxic chemicals must be placed in a ______________ landfill. To prevent the chemicals from escaping, the bottom and walls of this landfill are lined with ______ and a ______________________________

13. Describe two reasons why "secure" landfills must be monitored.
(1) ______________________________
(2) ______________________________

14. Do you think an area where granite has been mined would be suitable for a secure landfill? __________________ Why or why not? ________________

15. Would a desert be a good site for a secure landfill? __________
Give reasons to support your answer. ______________________________
____

16. Identify three factors that must be considered when selecting a site for a secure landfill. (1) ______________________ (2) ______________________
(3) ______________________

17. List three reasons why secure landfills are a very expensive means of waste disposal. (1) ______________________________
(2) ______________________________
(3) ______________________________

18. According to the EPA, good engineering and monitoring will **(eliminate the possibility of/reduce the risk of)** ground water contamination.

19. Would you object if land near your home was selected as a site for a secure landfill? _________ Why or why not? ______________________________
____

20. What are some possible alternatives of permanent storage of the hazardous waste?
____

Treat, Recycle, and Reduce Wastes [406]

21. Why did industries and municipalities choose landfills as the way to dispose of wastes? ______________________________

22. The waste water produced in paper making contains sulfur and carbon. The carbon is removed and heated with ____________ to make it "activated." The activated carbon is used in ________________. The sulfur water is recycled and used to ______________ more wood chips.

23. List two advantages of recycling waste water. (1) ________________________
(2) ______________________________

24. Experiments are being conducted to see if ______________ will digest hazardous wastes.

25. Some hazardous chemicals can be removed from water by using filters that contain particles of ______________________________. The chemicals cling to the surface or are ________________ by these particles.

Name ______________________________ Date ________ Class Period ____

26. Toxic metals that are dissolved in water may be removed by treating the water with chemicals which change the toxic metal to a ____________ which can then be removed with a ______________________.

Man-Made Rocks [406]

27. To build the Colosseum, workers mixed ________________ and ________ to make cement. Concrete made with this cement was very **(hard/soft)**.

28. Today the same process is used to make synthetic rocks that contain ____________________. The wastes are first mixed with ______ ash and then ________ is added.

29. How does this method of hazardous waste disposal decrease the possibility of water pollution? ____

30. The synthetic rocks may also be used in the construction of ____________.

31. Identify two industries that can use this technology to dispose of hazardous wastes.
____
____

32. Disposal in synthetic (man-made) rock can be used for _____ percent of hazardous waste. Describe two types of materials that cannot be stored in synthetic rock.
(1) ____
(2) ____

33. Explain why quality control is important when making the synthetic rock.
____
____

34. What types of tests are made by the EPA?
(1) ____
(2) ____

Incineration at Sea or on Land [407]

35. European countries destroy __________________ by incineration on specially-built ships.

36. The burning of wastes on incinerator ships:
( ) must meet standards set by the EPA
( ) must meet international standards
( ) is not regulated by any standards.

37. The major waste products produced when organochlorides are burned under the correct conditions are:
(1) ____________ (2) __________________ (3) ______________

38. What two conditions must be met to insure complete combustion?
(1) __________________________ (2) __________________________

39. What happens to the hydrochloric acid? ______________________________
____

Name ______________________________ Date ________ Class Period ____

40. Do you think that scrubbers should be required? __________ Explain.

____

____

41. According to EPA studies, will the burning of organochlorines have a long-term effect upon ocean life? ______

42. Why does the American Society of Mechanical Engineers recommend incineration of hazardous wastes? ____

____

43. Why are the environmental groups Alabamians for a Clean Environment and Friends of the Environment (Georgia) against incineration of hazardous wastes?

____

____

44. What do you think is the best way to handle the problem of hazardous waste?

____

____

Describe a situation where you would approve incineration as a means of hazardous waste disposal.

____

____

Introduction [410]

1. Some diseases, like typhoid fever, are caused by bacteria in water that has been contaminated with _______________.

2. In many countries diseases spread by contaminated water have almost disappeared. This has happened because of improved _______________ and _______________ of water supplies.

3. Most people in North America who lack a safe supply of drinking water live in:
   ___ large cities ___ small towns ___ rural areas.

4. About ________ percent of the wells and springs used by individual families in Pennsylvania are not safe to use without treatment.

5. Explain why Pennsylvania has more reported cases of water-related disease than other states. ____
   ____
   ____

6. In the developing world most sewage from urban areas is **(treated/untreated)** when it enters rivers.

7. Most people that die in developing countries are suffering from:
   ____ AIDS ____ cancers ____ diarrhea

8. In addition to lack of food, people in Ethiopia die from diseases like ___________ which are spread by contaminated water supplies.

9. Many areas of developing countries do not have a safe supply of drinking water. One of the goals of the United Nations is to provide a safe supply of drinking water to all nations. Do you think the UN will reach this goal? ________ Explain.
   ____
   ____

The Safe Drinking Water Act [411]

10. The Safe Drinking Water Act was passed in ___________. The law requires water suppliers to ___________ the water for substances that may _________________.

11. The **Maximum Contaminant Level** is the __________________________ of a substance that is allowed in the drinking water. The **MCL** of a substance is set by the ________. If tests show that the water contains more than the MCL of a substance the supplier must ____.

12. More than __________ chemicals have been found in drinking water supplies. By 1986 the EPA had set MCLs for _______ of these substances. Why hasn't the EPA set MCLs for all of these chemicals? ______________________________________
    ____

13. List three **inorganic contaminants** that have MCLs.
(1) ________________ (2) __________________ (3) _________________

14. List two types of **organic contaminants** that are regulated by MCLs.
(1) ____________________________ (2) ___________________________

15. Long-term exposure to some organic chemicals may increase the chance of ________________.

Organisms in the Water [411]

16. Has the EPA set MCLs for disease-causing bacteria? _____ for viruses? _____ If not, why? ____

17. What type of bacteria is an **indicator organism** for contamination of water supplies? ______________________. Where are these bacteria normally found? __________________________________ If a water supply contains Coliform bacteria, it means that the water has been contaminated with __________ ____________________

18. A water supply should not be considered safe to drink unless ________________ ____

19. If your water supply is a well or a spring, how often should it be tested? _______ _____________ Public water supplies must be tested each ____________. The number of samples is based on the __________________.

20. The number of bacteria counted in the samples taken from a public water system must average less than _____ per 100 ml of water.

21. Cryptosporidiosis, a disease caused by **(bacteria/parasitic protozoan)** in the water supply, causes ______________________ that may be life-threatening to ___________, ___________ people, and people with diseases such as ________.

22. *Giardia lamblia* is a **(bacteria/protozoan/virus)**. It is found in the intestines of ___________________________________.

23. **Giardiasis** is **(the most frequently/very rarely)** seen in the United States today. It affects the **(digestive/nervous)** system.

24. *Giardia* organisms survive **(a few months/only a few hours)** in the water.

25. *Giardia lamblia* is commonly found **(only in the southern states/ throughout the United States)**.

26. During a hiking trip in a designated National Wilderness Area, would it be safe to drink the water without first treating it? _____ Explain why. ________________ ____

Is Your Water Safe? [412]

27. The term used to describe water which is safe to drink is _________________. **(True/False)** It is best to assume that all water that has not been tested is contaminated and not safe for drinking unless treated.

28. Water from some wells is safe to drink without treatment. This depends upon the __________ and ________ of the soil as well as the __________________________ a source of pollution.

29. Even though water supplied by a water company meets EPA standards, it may be contaminated with __________. The source of lead is the ________________. The only way to determine if the water contains lead is __________________________.

30. Who needs to have water tested? ____
____

31. Well owners should have water tested for _____________ and ______________. The best time of year to test water for nitrates is the ____________. The test should be made **(each year/when the well is drilled)**.

More frequent tests are recommended if __________________________________ or if there are __________________________.

32. The EPA recommends that well owners have water tested at least once for ______ and __________. Depending upon the location of the well, tests for ____________ may be recommended.

33. Two factors that must be considered when selecting the location of the well and the septic system are: (1) ________________ and the (2) _____________________.

34. To prevent the well water from becoming contaminated by surface water:
(1) The steel casing should ____________________________________.
(2) The casing should be surrounded by ______________________________.
(3) The ground level should ____________________________________.

Making the Water Safe [413]

35. To determine the type of treatment needed, the water must be ______________.

36. The chemical used to disinfect most public and private surface water supplies is ____________________.

37. The amount of chlorine needed to treat the water is referred to as the ___________ ___________. The amount of chlorine needed can be affected by ______________ present in the water such as _____________________ or ____________.

38. The **contact time** recommended by the EPA is ________ minutes. The contact time may be reduced by increasing the ___________________________________.

39. If the contact time is 30 minutes, the amount of free chlorine in the water should be **(0.4 /4.0)** ppm. If the contact time is reduced to 5 minutes, an acceptable amount of free chlorine in the water is **(0.4 /4.0)** ppm.

40. The major objection to the use of chlorine is its __________. Despite this objection, chlorine is the most frequently used chemical because it is ____________________ and _______________________.

41. **(True/False)** It is easy to determine if water treated with chlorine is safe to use.

A simple test can determine the ________________________ or the amount of free chlorine remaining after treatment.

42. Identify the methods of treatment that effectively kill the *Giardia* parasite.
(1) ___________________________ (2) ______________________________

43. Why aren't *Giardia* parasites usually found in well water? __________________
____

44. **(Fluorescent/Ultraviolet)** light can be used to kill bacteria in drinking water. Before the water is exposed to the light, it must be ________________.

45. Compare chlorination and light treatment of water supplies. Place a **C** beside those features which are an advantage of chlorination and a **U** beside those features which are an advantage of ultraviolet light.

_____ no taste or odor _____ can be easily tested to see if water is safe
_____ least expense _____ time required to kill bacteria
_____ most reliable method to kill spores or cysts.

46. The water supplier must issue a Boil Water Advisory when:
(1) ____
(2) ____
(3) ____

47. If water contains *Giardia* parasites it can be used for drinking if it is _________ for at least ________ minute. Is it necessary to use treated water for making ice? ______ It is necessary to use treated water for washing dishes unless they are ________________.

48. An **activated carbon filter** may be needed to remove _____________ chemicals that cause __________, __________, or __________. If you live in a city with a population greater than 10,000, do you need to install an activated carbon filter in your home? _______

49. The best filter to use on the water system within a home is a _________________ filter that **(attaches to the faucet/is under the sink).**

50. Filters must be changed regularly to:
(1) __________________________ (2) ______________________________

51. **Turbid** refers to water that is **(cloudy/colored)**.

52. Particles that are not harmful may make water unfit to drink by ________________
____

53. **Turbidity** refers to the amount of __________ scattered or absorbed by the particles suspended in water. The MCL for turbidity is _____ TU.

54. Particles are removed by **(activated carbon filters/sediment filters)**.

55. Large water treatment systems use activated carbon filters to remove ____________ ________________ and graded sand filters to remove ___________________.

56. While large water systems use graded sand filters, single-user water systems might use a ____________________ filter.

57. Dairy cattle will not drink enough water if it contains high levels of ____________.

58. Well water containing high levels of iron is **(colorless/tasteless)**.

59. Iron dissolved in water changes into small red particles when the water is mixed with ____________ or _______________. These particles may be removed with a ____________________ filter.

Water Conditioners (Softeners) [416]

60. Identify three chemicals that can be removed from water by a water softener. (1) ________________ (2) ____________________ and (3) __________________

61. __________________ and __________________ decrease the efficiency of water heaters by forming deposits called ___________.

62. The person in charge of the cleaning and laundry may object to deposits of **scale** that dull the appearance of the _________________ and the ________________.

63. People who have ________________________________ should not drink water that has been treated in an ion-exchange water softener.

64. In an ion-exchange water softener, the calcium and magnesium ions are exchanged for ____________ ions in the **resin**.

65. When installing an ion-exchange water softener, many homeowners place the softener on the hot water line. Explain the advantages offered by:
A. Treating the hot water: ____
B. Not treating the cold water: ____
____

66. Identify three methods that can be used to remove nitrates from a water supply.
(1) ______________________________ (2) _______________________________
(3) ______________________________

What is the disadvantage of all of these removal methods? ____________________
____

In an Emergency [416]

67. It is safe to drink water that has been boiled _____ minutes.

68. The disinfectant that requires the longest contact time is **(chlorine/iodine)**.

69. Water purification tablets found at camping supply centers contain a form of **(chlorine/iodine)**.

70. **(Chlorine/Iodine)** should not be used for long periods of time because of its possible effects on ____________ children and people with _______________ problems.

Vocabulary Review

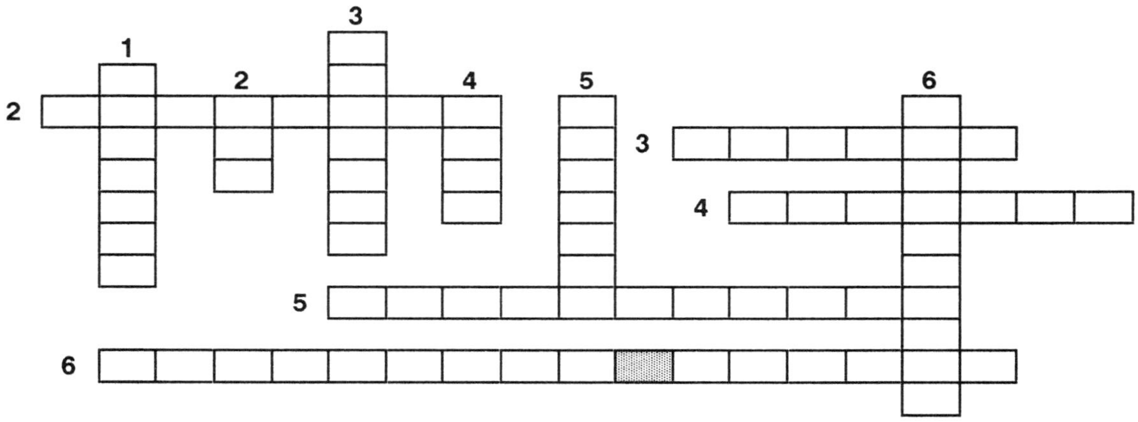

ACROSS:

1 chemical used to disinfect most water supplies.
2 an indicator organism which is used to detect animal waste.
3 water which is cloudy is:
4 the time required for a disinfectant to kill bacteria.
5 the type of light which is used to disinfect water supplies.
6 they type of carbon which is used in water filters.

DOWN:

1 word which indicates the water supply is safe to drink.
2 ____-exchange water softeners remove certain minerals from water.
3 disease spread by water supplies contaminated by human waste.
4 the amounts of the pollutants which the EPA allows in drinking water.
5 the most common cause of water-related disease in the U.S. today.
6 superchlorination or ________________ are effective ways to prevent giardiasis.

Name ______________________________ Date __________ Class Period ____

Unit IV — 3.4 The Waterworks

Introduction [419]

1. The Romans used canals or ____________________ to carry water from the mountains to the city.

2. The settlement of Bethlehem, Pennsylvania, was uphill from the spring which was the source of water. The first pump was powered by ____________ and the pipes were made of ________________________.

3. Like other cities, Bethlehem had to search for a new supply of water when the spring ____________________________. Today Bethlehem's water supply comes from ____.

4. Identify two major cities that obtain water from distant places.
(1) ______________________________ (2) ____________________________

5. What is the major source of water for your city? ___________________________

Have a Drink in New York City [419]

6. The major source of drinking water for New York City is ____________________
____

7. Scientists working for the water department include _________________________ and chemists.

8. Chemists test the water for traces of ____________________ and _____________ chemicals.

9. In the summer the reservoir is treated with __________________________ to kill **(algae/ bacteria)**.

At the Waterworks [420]

10. Water sold by both public and private water companies must meet criteria (standards) that are set by the _____________________________________.

11. Water entering the water treatment plant contains _____________ and other small organisms.

12. The purpose of primary chlorination is to ________________________________
____.

13. Aluminum sulfate or ____________ is added to cause the small particles to stick together. The clumps of particles are called ____________.

14. For proper floc formation the pH must be near ________. If it is necessary to adjust the pH, _____________ is added to the water.

Name ______________________________ Date ________ Class Period ____

15. Activated carbon is added to remove chemicals that cause ________, ________, or __________ problems. The amount of activated carbon required increases during the ______________ months.

16. ____________________________ may be added to remove organic chemicals including THMs. THMs (trihalomethanes) form when ______________ combines with organic chemicals.

17. Sodium zinc polyphosphate may be added to prevent ______________ of pipes. It forms a __________________________ on the inside of the pipes.

18. As water flows through the flocculation tank it must flow over and under a series of ____________. The purpose of these dividers is to **(increase the dissolved oxygen level/slow the speed)** of the water.

19. As the water moves though the flocculation tank, the floc grows __________ and heavier. Floc sinks to the bottom in the **(flocculation/settling)** tank.

20. Any floc not removed in the settling tank is removed by the ______________ filter. As the floc accumulates in the bed of sand, it _________ the flow of water.

21. The graded sand filter is cleaned **(with a high pressure water jet/ by reversing the flow of water)**. This process is called a ______________.

22. More backwashes are needed during the ______________ months because of the increased growth of ___________.

23. The wash water is pumped to a tank where the ____________ settle out. The water may be ______________ and treated for drinking water or returned to a ____________ or stream.

24. Filtration of the treated water reduces the risk of disease caused by ____________ ____________. If the waterworks does not have graded sand filters, high levels of ____________ must be added to kill these organisms.

25. Give the reason why these chemicals are added to the water before it leaves the waterworks.
chlorine: ____
lime: ____
fluoride: ____

26. Is fluoride added to the water supply in the city where you live? ____________

Name ______________________________ Date ________ Class Period ____

Unit IV — 3.5 Will There be Enough?

Introduction [425]

1. Three-fourths of the earth is covered by water, but most of this water is located in the ____________. The usable supply of fresh water is _______ percent of the total water supply.

2. Match the following volumes of water with the appropriate description.

The Earth's Water Supply	
___ A. glaciers and ice caps	(1) 1 inch (2.5 cm) cube
___ B. ground water	(2) 1/4 inch (0.6 cm) cube
___ C. surface water	(3) 1/8 inch (0.3 cm) cube
___ D. salt water (oceans)	(4) 1/16 inch (0.15 cm) cube

The Earth's Water Supply	—By the Drop
___ E. glaciers and ice caps	(1) < 1 drop
___ F. ground and surface water	(2) > 2 drops
___ G. salt water (oceans)	(3) > 97 drops

Distribution of Fresh Water	
___ H. glaciers and ice caps	(1) < 1 percent
___ I. ground water	(2) 14 percent
___ J. surface water	(3) 85 percent

The Distribution of Water [425]

3. The average annual rainfall in the United States is ______ inches. The average annual rainfall where you live is ____________ inches.

4. Applying what you have learned about weather, give two reasons why the distribution of rainfall is so uneven. [See Unit II Section 2.1 Weather Patterns.]
(1) ____
(2) ____

5. While the wettest place in the United States is in Hawaii, the region of the continental United States that receives the most rainfall is in the ________________ ________________. What geographic feature is responsible for so much rain being "dumped" in these two locations? ________________

6. West of the continental divide, most of the surface water comes from snow that falls on the ____________ Mountains. The major rivers that carry this water are the ______________ in the Southwest and the ______________ in the Northwest.

7. The major source of water for irrigation in the High Plains states, that lie east of the Rocky Mountains, is the ________________________.

8. Large amounts of water are needed for ____________________ and some ______________. Therefore water is often a ______________________ in the growth of a community.

Name ______________________ Date __________ Class Period ____

9. Many parts of the Northeast receive more than 40 inches of rainfall each year, but this is sometimes not enough to meet demands because of the high ______________________ density.

The Demand for Water [426]

10. Most people living in a poor tropical country probably use **(less than 5/more than 50)** gallons of water each day.

11. Many people living in the United States uses **(less than 5/more than 50)** gallons of water each day.

12. Do you think that the restrictions placed on water use during the drought in northern New Jersey allowed people enough water to meet their basic needs? ________

13. Which of the following has had the fastest rate of growth?
_____ population growth _____ demand for water

14. In the United States most water is used for ______________________. Farmers in California use about ______ percent of the water available and grow about ______ percent of the fruits and vegetables produced in the United States.

15. The second major use of water in the United States is the production of ___________________________.

16. The water used for the production of electricity at hydroelectric power plants and for transportation is used within the river. These uses of water are considered ____________________________ uses.

17. Identify three industries that use huge amounts of water. ____________________
____

18. Land-roving vehicles require large amounts of water yet water does not provide the energy for these vehicles. Explain their dependence upon water
____
____

19. In the western United States __________________________ is the number one user of water. In the East, __________ production uses the most water. Manufacturing uses more water in the _________.

20. The average amount of water used per person in many countries is far less than 100 gallons/day. The average amount of water used per person in the United States is 2,000 gallons/day. List three reasons why so much more water is used for each person in the United States.
(1) ____
(2) ____
(3) ____

The Mining of Water [428]

21. After resources beneath the land are removed, ____________________ or sinking of the land may occur.

Name ______________________________ **Date** ________ **Class Period** ____

22. ________________ refers to the removal of a natural substance from an ecosystem faster than it can be replaced.

23. The term mining may correctly be used to refer to the removal of:
____ coal ____ minerals ____ trees ____ water

24. Some people were forced to leave their homes in Baytown, Texas, because the ground was **(below sea level/unstable)** due to subsidence that occurred when _________ was mined.

25. Layers of porous rock filled with water are sometimes called underground reservoirs or _______________.

26. In southeastern Arizona water is mined for __________________. Mining of the water has caused __________________ of the land. This is causing problems for developers because the ground is now **(below sea level/unstable)**.

27. Approximately _____ percent of crops come from the _____ percent of the land that is irrigated. Since less than ten percent of farmland is irrigated with ground water, why is this farmland so important? ______________________________

____

28. The World Resources Institute estimated that water is being mined to supply _______________ of the irrigated crops.

29. Even though the aquifer will still contain water, officials estimate that many farmers in Texas will not be able to irrigate crops with it after the year 2000. Explain why.

____

____

30. The major source of water for irrigation in the High Plains states is the __________________ aquifer. It is a **(nonrenewable/renewable)** resource.

31. The High Plains states are important producers of ___________, _________, and ___________.

32. Within _____ years many farmers may not be able to get the water they need for irrigation from the Ogallala Aquifer.

Rivers and Water Rights [429]

33. Water that is not returned to the river is _______________. Some of this water becomes a part of a _______________, but much of it _________________.

34. **(Agriculture/Industry)** withdraws or consumes more than half of the water that it withdraws from ground or surface water.

35. Both the __________________ and ________________ Rivers which flow from Colorado have been rationed by the courts.

36. Most of the water in the Arkansas River that flows into Kansas is used for ________________________.

Name ______________________ Date __________ Class Period ____

37. How do courts determine who has the right to use the water in a river? __________
____

38. States in the upper and lower Colorado River Valley are each guaranteed 7.5 million acre-feet of water each year. Mexico is guaranteed _____ million acre-feet of water per year. Do you think that the United States should be obligated to send some of the Colorado River's flow to Mexico? ________

39. Southern California has been using more than its share of Colorado River water. Identify the two projects that will force southern California to give up some of this water.
(1) ____
(2) ____

40. Expensive canals may someday carry water from the _______________ River to western Kansas. Explain why the cost to transport this water will be high. ______
____

41. Identify two possible sources of the additional water needed by southern California.
(1) ____________________________ (2) ______________________________

42. Give two reasons why the demand for water is so great in southern California.
(1) ____________________________ (2) ______________________________

Vocabulary Review

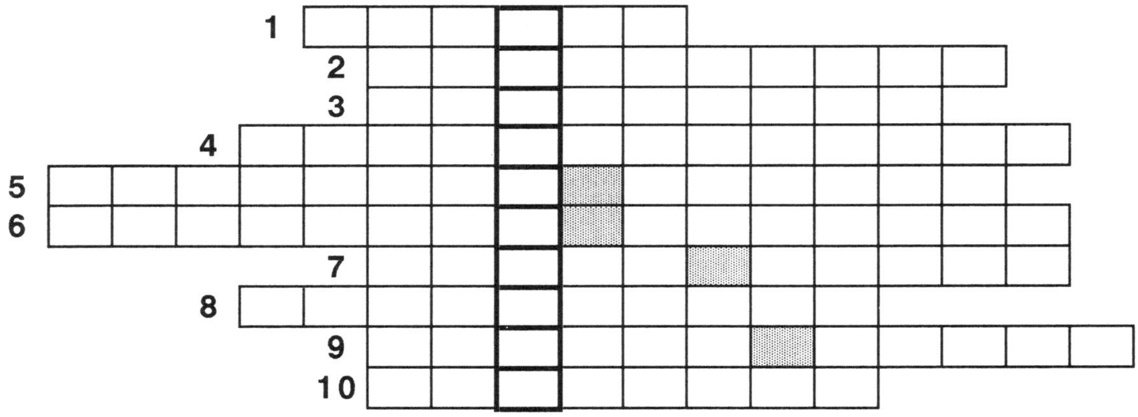

1 the removal of a natural resource from an ecosystem faster than it can be replaced.
2 most water which is consumed is used for _________________.
3 325,851 gallons or 1,200,000 liters of water is an ___________.
4 water which passes through turbines at a hydroelectric power plant is a _________________ use of water.
5 water is often a _________________ for the growth of a community.
6 ground water located beneath a large part of the High Plains States.
7 the highest level of water in an aquifer.
8 the sinking of land which occurs when resources beneath are removed.
9 nearly 1/4 of the all fresh water is _________________.
10 water which becomes a part of a products such as baby food is ___________.

Name ______________________________ **Date** ________ **Class Period** ____

Unit IV — 3.6 Making More (Water Available) by Using Less

Introduction [432]

1. Call an excavating contractor who advertises topsoil. What does a ton of topsoil cost? ________
   Call your water company. What does a ton of water cost? ____________ Is water really "cheaper than dirt"? _______

2. Do you think that we are a nation of water wasters? ________ Cite evidence that supports your position. ____

3. What factor will make water conservation a necessity? ______________________

Meters [432]

4. Which household will probably use water most efficiently?
   ___ a home that is assessed a flat rate.
   ___ a home that is assessed a frontage rate.
   ___ a home that has a water meter.

5. Which more accurately reflects the amount of water used — **frontage rate** or **flat rate**? _______________ Explain. ______________________________________

6. The method that determines the amount of water actually used is:
   ____ **flat rate** ____ **frontage rate** ____ **water meter**.

7. Water use **(decreased/increased/did not change)** after Boulder, Colorado, installed water meters.

8. Compare Phoenix and Tucson, Arizona.
   Which city uses more water per person?___________________
   Which city has the cheapest water rate?___________________
   Which city has lawns with desert plants?___________________
   In which city do industries recycle water? ___________________
   Based on the use of water in these two Arizona cities, how does the cost of water affect water use? ____

Leaky Water Systems [433]

9. Why don't cities with old leaky water pipes replace them? ___________________

10. Which saves more water—fixing a leaky faucet or fixing a leaky toilet? ________
    Describe how you can determine if a toilet has a slow leak. _________________

Name ______________________________ Date __________ Class Period ____

Water Saving Fixtures [433]

11. In addition to water, what other natural resource is saved when water-saving fixtures are installed? ____________________

12. What is the largest single user of water in the home? ______________________
List three changes in plumbing that will decrease the amount of water used.
(1) ____
(2) ____
(3) ____

13. Explain how reducing the amount of water used will improve the quality of the effluent that enters streams from the sewage treatment plants. ________________
____
____

A Change in Lifestyle [434]

14. Describe how water use can be reduced in each of the following activities:
A. washing dishes in the sink: ______________________________
B. washing dishes in a dishwasher: ______________________________
C. brushing your teeth: ______________________________
D. taking a shower: ______________________________

15. List two ways that restaurants can save water.
(1) ____
(2) ____

16. Explain why the water used to wash a china dish is "consumed," but the water to make a disposable dish is only "borrowed." ______________________________
____
____

17. Give one change the people of Phoenix, Arizona, could make that would drastically reduce their demand for water. ______________________________
____

18. Explain why **xeriscape landscaping** requires less water. __________________
____

Irrigation [435]

19. By improving the efficiency of irrigation, farmers can reduce the loss of ________ and ________________.

20. Computerized irrigation systems reduce the demand for water by:
(1) detecting ______________
(2) adjusting flow for ____________________ and ______________________

21. The most efficient irrigation system is a **(center-pivot/subsurface drip)** system.

22. What are two disadvantage of a subsurface system?
(1) ____________________________ (2) ____________________________

Name ______________________________ Date __________ Class Period ____

23. Why does a low-energy precision application (LEPA) irrigation system require less water than a center pivot irrigation system?

____

____

24. Give two ways that the farmer benefits from installation of an efficient irrigation system.
(1) ______________________________ (2) ______________________________

From Farm to City [435]

25. How can installation of subsurface or LEPA irrigation systems on farms in benefit people in cities? ______________________________

____

26. Why are cities in Arizona buying farms? ______________________________

____
What term refers to this method of obtaining water? ______________________________

27. How does "water ranching" affect the food supply? ______________________________

____

____

Name ______________________________ Date __________ Class Period ____

Unit IV — 3.7 Making More Fresh Water — Desalting the Sea

Introduction [437]

1. ________ percent of the world's water supply is in the sea.

2. The most abundant mineral dissolved in sea water is ______________________.

3. The minerals dissolved in sea water come from ___________ and ____________.

4. The process that increases the concentration of salts dissolved in sea water is ______________________.

How to ... [437]

5. Match the correct process with the statements below.

(1) electrodialysis **(3) reverse osmosis**
(2) flash distillation **(4) solar distillation**

___ A. ions are attracted to positive or negative electrodes.
___ B. water must be heated to 250°F (121°C).
___ C. pressure forces water through a membrane with small openings.
___ D. super-heated water passes through chambers with reduced pressure.
___ E. sea water evaporates because of the "greenhouse effect."
___ F. is not a good method for desalting sea water, but is effective method to remove salt from ground water with lower concentrations of salt.
___ G. fresh water condenses on a dome-shaped surface.
___ H. mineral ions move through large openings in the membrane toward positively or negatively charged rods.
___ I. fresh water is collected when steam condenses on pipes carrying the cooler sea water.
___ J. is the least expensive method of desalting water.

It Can be Done but ... [438]

6. What factor makes desalting sea water too expensive for most communities?
____

7. List one advantage and one disadvantage of solar distillation.
Advantage: ______________________________________
Disadvantage: ___________________________________

8. What is a less expensive water source for communities that can not afford to desalt sea water? ______________________________________

9. Explain why the sea is not likely to become a major source of fresh water for most regions that need additional water supplies for irrigation.
____
____

10. In addition to the cost of desalting, what additional problem must be solved when desalting large amounts of sea water? ____________________________

Introduction [439]

1. **(True/False)** In all states, if the label includes the word spring, the consumer can be sure that the water in the bottle came from a naturally flowing spring.
2. Before it is bottled, spring water is usually ____________ and _______________.
3. More than one-third of bottled water is taken from **(naturally flowing springs/city water supplies)** and further purified.
4. If water is labeled seltzer water or sparkling water, _____________________ has been added. ________________ are added to club soda.
5. **(True/False)** The FDA requires the label include the source of bottled water.

Water in a Bottle — Is It Safe? [440]

6. **(True/False)** Spring water is always pure because the water is filtered as it travels through the soil.
7. Water may become contaminated with high levels of certain ________________, leached from rocks and soil, and by disease-causing ________________ as it travels through the ground.
8. Before water from a spring in southern France is bottled and sold as Perrier, it is filtered to remove _____________. Bottles of Perrier were found to contain benzene after workers failed to ___________________________ on schedule.
9. **(True/False)** A consumer is usually correct when assuming that bottled water is safe to drink.
10. Bottled water that is bottled in one state and sold in another state is regulated by the ________________________________. If it is sold in the same state where it is bottled, water is regulated by the ___________________________.
11. **(True/False)** In some states, state regulations are more demanding than federal regulations.
12. **(True/False)** Bottled water is subject to more rigid water quality standards than tap water because it is only used for drinking.
13. Both bottled water companies and public water suppliers are required to test for ____________________ at least once each week.
14. **(True/False)** The water company is not required to put the results of water quality tests on the label unless the water does not meet the standards set for drinking water.

Water from the Tap — Is It Safe? [441]

15. **(True/False)** A homeowner can safely assume that tap water is safe to drink.

16. Bottled water may be the safest, least expensive alternative when tap water is contaminated with: (1) ____________________ (2) ____________________ (3)____________________________ (4) __________________________

17. To determine if bottled water is safer to drink than tap water, it is necessary to compare the ____.

18. **(True/False)** Information concerning the quality of water provided by a municipality or private water company should be available from the company.

19. **(True/False)** Information concerning the quality of water from a private well can be obtained from the state's department of environmental resources.

20. **(True/False)** The only reliable way to determine if well water is safe to drink is to have the water tested.

Is It Healthier? [441]

21. **(True/False)** Drinking bottled water is a good way to obtain calcium and other minerals needed in the daily diet.

22. **(True/False)** According to the AMA, most consumers would probably be healthier if they spent the money for exercise classes at a health club instead of buying bottled water.

23. **(True/False)** If your lifestyle requires that you drink something from a bottle, bottled water is a healthy alternative to beer or soda.

24. **(True/False)** People living in New York City must drink bottled water to avoid the high levels of lead in the city's water supply.

25. **(True/False)** Some city water supplies are healthier for you than some brands of bottled water.

Does it Taste Better? [442]

26. Most bottled water is purified with ____________. Most city water supplies are purified with ______________. Both of these chemicals kill ________________.

27. What is the advantage of chlorine? ___________________________________
What is the disadvantage of using chlorine? ______________________________

28. The taste of water that has not been treated is due to _______________ in the water. The only water that lacks minerals is _________________ water.

29. **(True/False)** Distilled water is usually purchased for drinking.

30. Water from some deep wells has a high level of ___________ which makes it unacceptable for drinking.

31. Under what circumstances would you buy bottled water rather than drinking water from the tap? ____
____

Unit IV – 3.9 Fluoride in Your Water — How Much is Enough?

Introduction [444]

1. **Colorado Brown Stain** or **Texas Teeth** refers to teeth that are permanently _______________ from exposure to ___________________.

2. Fluorides are found in most __________, __________, and __________. The region of the U.S. where water has high levels of fluoride is the.______________.

Scientific Studies [444]

3. The subjects of the first scientifically controlled studies that tested the effects of fluoride were ________. The diet of animals in the control group **(did/did not)** contain added fluoride.

4. The 10-year study of humans showed that people who drank water with ________ ppm fluoride from infancy until they were _________ years old had fewer cavities.

Fluoride — How it Works [444]

5. The part of the tooth above the gum line is covered by ____________, the hardest substance in the body.

6. When bacteria digest the sugar in the food we eat they produce __________.

7. A cavity is formed when the _______________ in the tooth enamel are dissolved by the acids faster than the enamel can be rebuilt.

8. Scientists think that fluoride helps prevent cavities by:
   A. reducing the bacteria's ability to digest ___________ and produce ________.
   B. making the _____________ more resistant to the action of the acid.
   C. increasing the _________________ of the enamel.

Fluoridation of Water Supplies [445]

9. Studies showed that the level of decay was 50–60 percent less in areas that had _____ ppm fluoride.

10. Results of the study begun in 1945 showed increasing the level of fluoride in water to ________ could reduce the number of decayed, missing, or filled permanent teeth by ________ percent.

11. More than _______ percent of the people living in the U.S. drink water supplies which are fluoridated. The level of fluoride is about ___________.

12. EPA regulations prohibit drinking water from containing more than ______ ppm fluoride. If the level of fluoride is higher, some of the fluoride must be _________.

13. Slightly more fluoride is added to water supplies in cities with **(cold/warm)** climates. This is based on the assumption that people in warm climates will ______________________________________.

14. **(Children/Teenagers/Adults)** receive the greatest benefits from drinking fluoridated water.

Too Much of a Good Thing — Fluorosis [446]

15. Industrial processing of phosphate rock that contains fluorides has resulted in the death of: ___ animals ___ humans ___ plants.

16. Only 0.1 ppb fluoride causes the death of certain __________ species. Other plants, including ______________ change fluoride into a harmless form. This allows high levels of fluoride to enter the _______________.

17. Symptoms of fluoride poisoning, or _________________, in dairy cattle would include: (1) _____________________________ (2) ____________________

18. Identify the condition and describe the changes that occur in human teeth in areas where the concentration of fluorides in the water supply is:
> 1.5 ppm ____
> 4.0 ppm ____

19. Describe the symptoms of **skeletal fluorosis.** ____ This condition is seen in people who have been drinking water with more than ______ ppm fluoride. This condition is common in some regions of **(Asia/Africa/United States).**

20. In addition to fluoride, the risk of skeletal fluorosis is affected by the intake of vitamin ______, ___________, and minerals ____________ and _____________.

The Anti-fluoridation Campaign [447]

21. The Public Health Service goal of 95 percent fluoridation of approved public water supplies by 1990 **(was/was not)** met.

22. In spite of the fact that ____ national health and science organizations support fluoridation of water supplies, only ____ states require fluoridation of approved public water supplies.

23. **(True/False)** Adding fluoride to the water supply caused illness among residents of Newburgh, New York.

24. **(True/False)** Studies in which animals were given fluoride revealed a possible link between high doses of fluoride and bone cancer, birth defects and infertility in some animal species.

25. **(True/False)** Studies of humans who live in areas that have high levels of fluoride in the drinking water reveal a link between fluoride and bone cancer.

26. **(True/False)** There is evidence that fluoride in the drinking water may cause kidney disease and birth defects.

27. State the position of the Coalition for Pure Water on the issue of fluoridation of public water supplies. ____
____

28. Under what circumstances will babies take in more fluoride than is recommended by the American Dental Association? ________________________________

____

Other Ways to Get Fluoride [448]

29. FACT: If fluoride is available during the first _______ years of life the tooth enamel will be more _______________ to acids.

FACT: The most effective way to provide fluoride to every one is to __________

____

FACT: The least expensive way to provide fluoride to every one is to __________

____

FACT: The American Dental Association says that children in communities with fluoridated water have ________ percent fewer cavities than children living in communities without fluoridated water.

30. **(True/False)** The American Dental Association says that using toothpaste with fluoride is effective in preventing tooth decay.

31. **(True/False)** Fluoride mouth rinses have been proven effective.

32. Should children under 5 years of age be given fluoride mouth rinses? _____

33. Should children under 5 years of age brush their teeth with fluoride tooth paste?

____

34. Daily swallowing of fluoride mouth rinses or toothpaste could lead to __________ _______________.

35. Should babies who live in areas with fluoridated water supplies be given vitamin drops with fluoride? ____

36. Dental fluorosis occurs when children get too much fluoride before the age of ____.

37. Why is it difficult to determine a person's exposure to fluoride? _____________

____

38. What is the National Research Council's recommendation concerning the risk of exposure to low levels of fluoride? ______________________________

____

39. Is your community water supply fluoridated? __________ Do you feel that it should be fluoridated? __________ Explain why you feel this way. __________

____

____

Name ______________________________ **Date** ________ **Class Period** ____

Unit V — 1.1 Energy — The Rest of the Story

Introduction [453]

1. List below the foods that were served the last time you ate a complete meal.

   lemon chicken sugar snap peas mushrooms noodles

   The following types of energy may have been used in the <u>production</u> of the food, for <u>storage</u>, for <u>transportation</u>, or to <u>prepare</u> and <u>serve</u> it. For each type of energy, give one example that shows how it may have been used.

   Coal: ______________________________
   Natural gas: ______________________________
   Nuclear: ______________________________
   Oil: ______________________________
   Solar: ______________________________
   Water: ______________________________
   Wind: ______________________________

Its the Law [453]

2. **Energy** is the ability to do ________, and ________ is the movement of matter.

3. The **Law of Conservation of Energy** says that energy can not be__________ or ______________, but it can be changed from one __________ to another.

4. The total amount of energy in the universe **(decreases with time/never changes)**.

5. Give an example of each of the following.
   Light energy is changed into chemical energy. ______________________
   Chemical energy is changed into heat energy. ______________________
   Heat energy is changed into mechanical energy. ______________________
   Mechanical energy is changed into electrical energy. ______________________

6. When energy is changed from one form to another, the amount of usable energy **(decreases/increases/remains the same)**.

7. When energy is changed from one form to another, the energy that can not be used is usually in the form of ______________.

8. Energy **(can/can not)** be recycled.

9. Energy efficiency is a measure of the amount of energy which does __________ ______________.

10. In each of the following pairs, identify the invention which is the most energy efficient.
    A. natural gas furnace/ oil furnace
    B. gas water heater/ electric water heater
    C. steam locomotive/ diesel locomotive
    D. incandescent lamp/ fluorescent lamp
    E. natural gas furnace (high efficiency)/ wood stove (high efficiency)

Name ______________________________ **Date** ________ **Class Period** ____

A Little History [454]

11. In early societies energy for work was provided by __________ or __________.

12. The first product of technology that reduced the need for human labor was the ______________.

13. Moving water provided the energy needed to power machines which ________ grain, _______ ore, _______ air, _____ wood, and even _________ clothes.

14. The major source of energy in Europe during the 13th century was __________. A shortage of this energy supply led to the use of __________.

15. ______________ and ______________ were the major sources of energy for Europe's industrial revolution.

16. The location of factories and cities depends upon the presence of a good supply of water and __________________.

17. **(Waterwheels/Windmills)** were an important source of energy in early America.

18. Identify the major source of energy used for each of the following activities in 1776.
Heating homes _________ Travel _____________
Cooking __________ Manufacturing a cannon ____________

19. The type of energy used is often determined by the cost. When wood became expensive, people began heating homes with __________.

20. The steam engine was first used to supply the energy needed to remove **(coal/water)** from the coal mines. The source of energy used to produce the steam was ___________.

21. The first steam boats had boilers which burned ____________.

22. In 1868 the products of technology which supplied most of the power used by industries were __________________ and __________________.

23. After the windmill was invented, ________ years would pass before inventors produced another source of power — the steam engine.

24. By 1900 ____________________ had become the major source of power for industry. The engines which powered industry in the 1900s were smaller but much more powerful than the first engines. The source of energy for this new type of steam engine was __________.

25. Identify the invention which supplied energy for each of the following:
threshing wheat ________________ pumping water ________________

26. Before World War I most of the energy on the farm was supplied by __________ and ______________.

Name ______________________________ Date __________ Class Period ____

27. List two reasons why fewer farmers were needed once the horses had been replaced by tractors.
(1) ____________________
(2) ____________________

28. Modern industry and agriculture are dependent upon large supplies of __________ and the products of ____________________.

29. List 4 ways in which farmers are dependent upon off-the-farm energy sources.
(1) ____________________
(2) ____________________
(3) ____________________
(4) ____________________

30. A society where people live in one community and work in another can exist only where there is an ____________________ and _____________ supply of energy.

31. It has been estimated that people living in the suburbs use 42 percent more energy than people living in the city. List as many reasons as you can why the demand for energy is greater in the suburbs than the city.
(1) ____________________
(2) ____________________
(3) ____________________
(4) ____________________

32. President Truman's administration predicted that we would:
( ) develop the technology for renewable energy sources by the 1970s.
( ) become dependent upon oil from the Middle East by the 1970s.
( ) have a major accident at a nuclear power plant during the 1970s.

33. By the 1970s the U.S. had only _____ percent of the world's population, and used more than _____ percent of the total energy consumed by all countries.

34. By the 1990s the U.S. had only _____ percent of the world's population, and used _____ percent of the total energy consumed by all countries.

35. Embargo — a government order prohibiting or restricting transportation of a product. In the 1970s, an embargo was placed on _______ exported from the Arab countries. Predict the effects of the embargo on a country that depends upon foreign oil.

____
____
____

Name ______________________________ Date ________ Class Period ____

Unit V — 1.2 Fossil Fuels — Energy for a Nation

Introduction [457]

1. During the 1970s there was a shortage of ____________ and ______________.

2. During the 1970s there was a significant **(decrease/increase)** in the price of crude oil.

3. In the 1970s:
The Energy ________________ and ________________ Administration was established.
The Energy Policy and ______________________ Act was passed by Congress.
The ______________ Heating and Cooling Demonstration Act was passed.
Congress passed a law that reduced the speed limit on all interstate highways to ________ mph.
The federal government promoted ___________________________ technology.
Manufacturers were required to produce cars with improved _______________
Manufacturers were required to provide **EnergyGuides** that allowed consumers to ____ of different models.
Homeowners who made their homes more ________________________ were given a tax break.

4. During the 1980s Congress passed a law that allowed states to **(decrease/increase)** the speed limit on interstate highways.

The Changing American Scene [457]

5. The major sources of energy used in 1850 were ___________, ____________, and _________________. Place a **R** beside each source that is renewable.

6. Which source of energy listed above was used at a rate that made it a nonrenewable resource? ____________

7. _________________, ___________________, and ______________ are the major sources of energy used today. Are these sources renewable? ________

8. The U.S. began importing oil in 19 _____. By 1973 ________ percent of the oil used in the U.S. was imported. Ten years later, in 1983, imported oil provided only _______ percent of the energy used in the U.S.

9. During the **Arab Oil** _________________ the U.S. received no oil from the Arab States. This forced the government to make ______________ changes that ________________ our dependence on foreign oil.

10. The country which imports more oil than any other is **(China/Japan/Russia/the United States)**.

11. The U.S. imported ________ million barrels of oil per day in 1990.
If 20 gallons of gasoline can be produced from each barrel, how many gallons can of gasoline can be produced from the oil imported in one day? _________ million gallons of gasoline

If an average car holds 15 gallons of gasoline, how many car owners could fill their gas tank each day from the supply made from imported oil? _________ million

12. Most of the world's proven reserves are in the area surrounding the **(Gulf of Mexico/Persian Gulf)**.

It Began a Long Time Ago [458]

13. It takes many years for decomposers to partially break down the dead plants in a swamp. The partially decayed material is called _____________

14. If the plant material is buried deep enough to provide great amounts of __________ and ________________, the peat is changed into _____________.

15. Unlike most other rocks, coal burns because of ______________________ in the partially decomposed plant material.

16. Match the statements below with the correct type of coal.

(1) anthracite **(2) bituminous** **(3) lignite** **(4) subbituminous**

_____ A. is a very soft coal that crumbles easily.
_____ B. Large deposits of this type of coal are found in western states.
_____ C. has the highest carbon content.
_____ D. Most of the coal found in the United States is:
_____ E. has the lowest heat content.
_____ F. with 40–60 % carbon, it has a higher heat content than lignite.
_____ G. Nearly all of this type of coal is found in Pennsylvania.
_____ H. often called soft coal.
_____ I. hard coal.

17. Coal is formed from plants that once grew in _________________. Oil and gas are formed from plants and animals that once lived in ___________________________

18. Great amounts of heat and pressure over millions of years changed the organic matter into a mixture of hydrocarbons that includes __________________________ and ______________________________.

19. Match the type of rock with the correct description.

(1) reservoir **(2) source** **(3) trap**

______ A. rock that contains a large amount of organic matter.
______ B. rock that contains pores where oil and gas are stored.
______ C. rock that forms a dam or lid to prevent the escape of oil or gas.

20. _______________________ scientists are hired by oil companies to find rock formations that contain gas and oil deposits.

21. To find oil or gas deposits, scientists study the physical properties of the rocks. Physical properties measured include:
(1) __________________ (2)________________ (3) ___________________

Name ______________________________________ **Date** _________ **Class Period** ____

22. ________________________ a deposit that can be economically be recovered with present technology.

23. In an area with known gas deposits only one of every _____ wells drilled will produce gas.

Natural Gas [460]

24. The region with the largest natural gas reserves is the **(United States/Persian Gulf/countries of the former Soviet Union/South America)**.

25. In the United States, most gas reserves are located in states bordering the **(Atlantic/Gulf/Pacific)** Coast.

26. Natural gas supplies nearly **(1/4, 1/2, 1/3)** of our energy needs.

27. The United States imports natural gas by pipeline from ________________. Natural gas may be imported from other countries in tankers as LNG or ________________ natural gas.

28. Describe how natural gas is used to produce each of the following:
Bread: ____
Clothing: ____
Newspapers: ____
Furniture: ____
Automobile: ____

29. Companies use of natural gas to produce electricity and fuel fleets of vehicles to meet ________________________ standards without expensive ________ ______________ controls.

30. The major hydrocarbon in natural gas is _________________.

31. Natural gas is a clean-burning fuel because sulfur and other _________________ are removed before sending it to the customer.

32. Rank these forms of energy beginning with the form that has the most energy per unit mass. _____ Coal _____ Natural Gas _____ Oil

33. It is least expensive to transport energy in the form of **(coal/natural gas/oil)**.

34. The form of energy listed in question 32 which is the most expensive to transport is ___________________.

35. The most efficient and least expensive means of transporting natural gas over land is **(insulated tank trucks/pipelines)**.

36. Transporting natural gas by truck or ship is much more expensive than transporting oil by these means. Explain why the cost of transporting natural gas is greater.
____

Name ______________________________ Date ________ Class Period ____

37. Give one advantage of using natural gas to:
Heat homes: ______________________________
Cook: ______________________________
Industrial processes: ______________________________

38. Before processing natural gas, it **(does/does not)** have an odor.

39. There is a **(limited/unlimited)** supply of natural gas. If we continue to use natural gas at the current rate and did not find more reserves, the natural gas supply would last __________ years.

Petroleum (Crude Oil) [461]

40. Two-thirds of the world's proven oil reserves are located in the countries **(surrounding the Persian Gulf/of the former Soviet Union)**.

41. About ______ percent of energy used in the U.S. is supplied by oil. More than _____ percent of the oil we use is imported.

42. The demand for oil is **(decreasing/increasing)**.

43. List 5 products you use that are made using chemicals which come from crude oil.
(1) ______________, (2) ______________, (3)______________,
(4)______________, (5)______________

44. **Secondary recovery methods** allow recovery of **(additional oil/natural gas)** from oil wells.

45. List four types of chemicals that are pumped or injected into oil wells to recover additional oil. (1) ____________ (2) ____________ (3) ____________ ____________ (4) ____________.

46. List two physical processes which are used to recover additional oil.
(1) ______________ (2) ______________

47. Give one advantage and one disadvantage of the secondary recovery method.
ADVANTAGE: ______________________________
DISADVANTAGE ______________________________

48. Chemicals in crude oil are separated by the process of ______________.

49. ______________ refers to the processes involved in the separation of the chemicals in crude oil and the production of products from these chemicals.

50. ______________ are used to increase the speed of the chemical reactions during the manufacture of petroleum products including gasoline.

51. Rank the methods used to transport oil and other chemicals from most energy efficient to least energy efficient.
(1) ____________ (2) ____________ (3) ____________ (4) ____________
Most Efficient -- Least Efficient

52. Trucks are most often used to transport oil over **(long/short)** distances.

Name ______________________________ Date __________ Class Period ____

53. Describe two ways that oil spills affect the environment.
(1) ____
(2) ____

54. Greater amounts of pollution are produced by the burning of **(coal/oil)**.

55. **(Automobiles/Industry)** produce most of the air pollution which is created by the burning of oil.

56. Indicate whether each of the following statements is **true T** or **false F**.

___ A. Scientists expect to find major oil reserves which have yet to be discovered.
___ B. Production of oil in the U.S. is declining.
___ C. On a worldwide basis production of oil is increasing.
___ D. On a worldwide basis discovery of oil is increasing.
___ E. The chance of finding a wildcat well that is productive is less than 10 percent.

57. The results of ______________________ testing indicated that there was a large reservoir of oil in an unexplored area near Prude Hole Bay oil fields. But rock formation did not have a ____________ and the oil had escaped.

58. Do you think that oil companies should be permitted to drill for oil and gas in National parks and National Wilderness Areas? _________
Give reasons to support your answer. ______________________________
____
____

59. When production declines an oil company must decide whether to cap the well or use secondary recovery methods. What is the basis for this decision? __________
____

60. As the price of oil increases, many industries and power plants use ____________ instead of oil as a source of fuel.

Coal [464]

61. Coal is the **(least/most)** abundant fossil fuel.

62. The largest exporter of coal is **(China/the countries of the former Soviet Union/the United States)**.

63. In 1990 coal supplied nearly **(1/4, 1/3, 1/2)** of our energy needs.

64. Most coal was used for:
___ producing electricity ___ manufacturing ___ heating homes

Name ______________________________ **Date** ________ **Class Period** ____

65. Identify the type of coal used for each of the following processes.

(1) anthracite (2) bituminous (3) lignite/subbituminous

_____ A. used by most steam-turbine power plants to produce electricity.
_____ B. mainly used for heating homes.
_____ C. use is limited to manufacturing/production of electricity.
_____ D. used to heat commercial buildings and factories.
_____ E. used to produce coke for the steel industry.

66. Identify the two conditions which are needed to change coal into **coke.**
(1)________________________ (2) ________________________

67. Unlike coal, coke is nearly pure ____________. It is burned with limestone and iron ore to produce _________ for making _________.

68. _________________ is a by-product produced during the making of coke and is used by farmers as a fertilizer.

69. Identify two products you use that are a by-product of the process of making coke.
(1) ________________________ (2) ________________________

70. Identify the type of mining, Surface mining **S** or Underground mining **U**, described by each of the following statements.

_____ **most coal recovered** _____ **strip** _____ **safest**
_____ **requires reclamation** _____ **longwall**
_____ **room and pillar system** _____ **least expense**

71. Explain why a company that is producing coal by surface mining would hire a biologist.

____

____

72. Abandoned strip pits are being reclaimed with money from:
( ) **landowners** ( ) **mining companies** ( ) **tax on coal**

73. Indicate the type of underground mining which the following statements describe.
room and pillar R or **longwall L**

___ most common type of deep mining.
___ roof is allowed to collapse.
___ used in very deep mines.
___ roof held in place by movable supports.
___ more coal can be recovered using this system.
___ columns of coal are left to support the roof.

74. At the processing plant, coal must be ____________, ____________, and _________________.

75. Bituminous coal often contains large amounts of ______________ which can not be removed by washing.

Name ______________________________ Date _________ Class Period ____

76. Although pipelines are often the most efficient way to transport gas or oil, they are seldom used to ship coal. In some locations, pipelines are not used because of a lack of ____________.

77. The cheapest way to ship coal over long distances is by **(land/water)**. When shipped by land the most efficient means is by ________________.

78. Some electrical power plants east of the Mississippi River import coal from Montana because it contains **(less sulfur/more energy)**.

79. The leading cause of work-related death for coal miners is **(accidents/lung disease/explosion and fires)**.

80. Describe how each of the following changes, brought about by the **Coal Mine Health & Safety Act**, reduced the death rate of coal miners.
Training reduces ____________________.
Scientific plans help prevent ______________________.
Detectors prevent death due to a build up of ______________________.
Ventilation prevents the formation of too much ____________________.
Rock dusting: ____

81. When water is released from coal mines without treatment, the high level of ______ may cause the death of aquatic organisms.

82. The respiratory system may be damaged by breathing air pollution which results from burning coal that contains _____________. Pollution control devices must be used if the coal contains more than _______ percent of this pollutant.

83. Identify the problem associated with abandoned coal mines which forced people from their homes in:
Pittsburgh, PA ____________________ Centralia, PA _________________

84. Coal will become **(less/more)** important as a source of energy in the future.

Synthetic Fuels from Fossil Fuels [467]

85. During the 1800s street lights burned "town gas" which was made from ________. Town gas was replaced by _____________________.

86. List three reasons why natural gas replaced coal gas.
(1) _________________ (2) _________________ (3) _________________

87. Much of the energy needed by Germany to fight World War II was supplied by synthetic fuel made from ___________. After the war, liquid coal was replaced by __________.

88. **Synthetic fuels** are made by chemically changing _______________________ or other _____________ substances. Synfuels are similar to _________________ products and _____________ gas.

89. Fossil fuels that are used to make **synfuels** are _________, ________________, and ____________________.

Name ______________________________ **Date** ________ **Class Period** ____

90. Synthetic oil is made from coal by mixing the coal with a ______________ and adding ______________ while it is heated to high temperatures under ______________.

91. One ton of coal will produce _______ barrels of synthetic oil. The synthetic oil contains only _______ percent of the energy in the coal.

92. In 1990, the U.S. imported 9.7 million barrels of crude oil each day. Assuming that a barrel of synthetic oil would produce the same products as a barrel of crude oil, how many tons of coal would we need to mine each day to produce the oil imported in 1990? _______ million tons

93. Modern technology can produce a high quality gas from coal. The coal is heated with ____________ and ____________. After impurities are removed, ____________ is added. Chemical reactions between the gases produces ____________ which is the major gas in natural gas.

94. Explain why synthetic gas made from coal can not be used to heat homes.

____

95. To remove oil from oil shales or tar from tar sands, the rock must be __________, ______________, and ______________.

96. One ton of oil shale will produce _____ gallons of oil. In 1990 the U.S. imported 7.3 million barrels of crude oil each day. [A barrel of oil = 42 gallons/189 liters]

Assuming that a barrel of synthetic oil from oil shale would produce the same products as a barrel of crude oil, how many tons of oil shale would need to be mined each day to produce the oil imported in 1990? _______ million tons

_____ barrels of water are needed for each barrel of synthetic oil made from oil shale. How many barrels of water would be needed to produce synthetic oil to replace the oil imported in 1990? ________ million barrels.

97. Compare the cost of petroleum products made from synfuels with the cost of petroleum products made from crude oil. Which will be more expensive?

____

98. Identify two environmental problems that may occur during the production of large amounts of synthetic oil from oil shale.
(1) __________________________ (2) __________________________

99. The production and use of synthetic oil will **(decrease/increase/not change)** the level of carbon dioxide in the atmosphere.

100. The production of oil from oil shale will require thousands of workers. Oil shale deposits are located in a region of the U.S. which has a low population density. What environmental changes may result from the increasing population?

____
____
____

Name ________________________________ Date _________ Class Period ____

Unit V — 1.3 Achieving Energy Independence

Introduction [472]

1. **(True/False)** Because of "Project Independence" we are no longer dependent upon foreign oil.

2. In the 1970s more than _______ percent of the energy used on the Hawaiian Island was imported. The major energy source imported is **(coal/natural gas/crude oil)**. To reduce its dependence upon imported energy, Hawaii began to develop ____________________ energy sources.

Thank You Thomas Edison [472]

3. A _______________ has large curved blades that rotate like the blades of a fan. It is connected by a rotating shaft to a __________________.

4. A **generator** consists of coils of _____________ wire and an electromagnet which is _______________. The movement creates a magnetic field which causes _________________ to flow along the wire producing ____________________.

5. Like many modern-day power plants, Thomas Edison's electric generating station produced electricity by burning _________ to produce ________ energy in the form of _________. The **turbine** converted this energy to ________________ energy. The **generator** converted this energy to ____________________ energy.

6. Electricity is a secondary source of energy. In 1991 ______________________ provided the primary source of energy that produce ______ percent of all electricity generated in the U.S. Power plants which provided the remaining 32 percent of electricity were either ______________ (22%) or ___________________ (10%).

Hydroelectric Power [473]

7. The primary source of energy that turns the turbines at hydroelectric power plants is ___________________.

8. Compared to the early 1900s, **(less/more)** of the electricity produced today is being produced by hydroelectric power plants.

9. Electricity produced at hydroelectric power plants is usually **(less/more)** expensive than electricity produced at nuclear power plants.

10. Production of electricity at hydroelectric power plants reduces _______ and ___________ pollution.

11. What is the major disadvantage of the production of hydroelectric power?

    ____

12. Hydroelectric power **(is/is not)** a renewable resource.

13. The amount of electricity produced at hydroelectric power plants **(is/is not)** expected to increase significantly.

Name ______________________________ Date ________ Class Period ____

14. Identify the three sites where additional hydroelectric power plants may be built in the future.
(1) ______________________________
(2) ______________________________
(3) ______________________________

15. The U.S. **(does/does not)** import electrical energy. If electrical energy is imported, identify the country which exports it. ________________

16. The amount of electricity produced by moving water depends upon the volume or __________ of water and its __________ or the distance it falls.

17. Why do land owners who have a stream flowing across their property need a permit to build a hydroelectric power system? ______________________________
____

Nuclear Power [474]

18. The major fuel used by nuclear power plants is __________________.

19. Energy is released from the fuel when the atoms split apart. This process, called ____________, releases huge amounts of __________ energy which is used to produce ___________.

20. At nuclear power plants and power plants that burn fossil fuels, turbines are turned by ___________.

21. Nuclear power plants produce electricity without producing __________ and ___________ oxides.

22. Under normal operating conditions, nuclear power plants release **(fewer/more)** radioactive substances into the environment than coal-fired power plants.

23. List three reasons why fewer nuclear power plants are being built.
(1) ______________________ (2) ______________________
(3) ______________________________

24. Identify the events that have caused concern about the safety of nuclear power and have led some countries to establish a "no nuclear" policy. ________________
____

25. Current plans are for nuclear energy to provide the major source of electricity in **(France/Sweden/the United States)**. Other countries that are continuing to build nuclear reactors are __________________ and ___________.

26. **Spent fuel** from a nuclear power plant **(is/is not)** radioactive. Although electricity has been generated at nuclear power plants for nearly 40 years, there is no permanent site for storage of spent fuel. **(True/False).**

Name ______________________________ **Date** ________ **Class Period** ____

Geothermal Energy [476]

27. Some geothermal energy was trapped when the earth was formed. Identify the three processes which continue to produce geothermal energy.
(1) ______________________ (2) ______________________
(3) ____

28. In most states geothermal energy will not be a major energy source because:
_____ **not enough heat is produced** _____ **the heat is too deep.**

29. List several ways in which geothermal energy is used:
(1) ______________________ (2) ______________________
(3) ______________________ (4) ______________________

30. List three problems that are associated with geothermal energy.
(1) ____
(2) ____
(3) ____

31. Is geothermal energy a renewable resource? ______________________
____

32. Identify the four <u>states</u> which have the potential to produce geothermal energy.
(1) __________ (2) __________ (3) __________ (4) __________

33. Development of power plants that use geothermal energy could **(eliminate/ reduce)** the need for foreign oil in these states.

Solar Energy [477]

34. Explain how each of the following forms of renewable energy depends upon the sun.
Wood ____
Wind ____
Water ____

35. Describe two features which must be included in the design of a home that will allow passive solar heating in the winter while avoiding excessive heating in the summer.
(1) ____
(2) ____

36. Objects that absorb and store heat energy are referred to as ______________.
Give four examples. (1) ____________ (2) ____________
(3) ____________ (4) ____________

37. A **passive solar collector** contains ______________ which is heated by solar energy. Cool air enters the collector through a vent near the __________ of the collector. It is warmed by heat energy from the thermal mass in the collector and leaves the collector through vents at the ________ of the collector.

38. An **active solar system** requires the use of a __________ or ________ to circulate the air or water in the system.

Name ______________________________ Date ________ Class Period ____

39. The most cost-effective active solar systems are used for **(domestic water/ space)** heating.

40. **Photovoltaic cells** change sunlight into ____________________ when the sunlight strikes ______________ in the conductor and releases electrons.

41. Give two disadvantages of using photovoltaic cells to produce electricity. (1) ________________ (2) ____

42. What is the major disadvantage of using solar panels to heat water in an electrical power plant? ____

43. In the 1990s solar energy **(will/will not)** become a major source of energy for the production of electricity.

Wind Machines [479]

44. At one time windmills were an important source of energy on farms. They provided energy for (1) _____________________ and (2) _________________________

43. Energy supplied by the giant windmill in Vermont was replaced with energy from ___________ and _________. These sources of energy were cheap and often more __________________ than wind.

44. Energy supplied by the wind is free, but the use of wind power has several disadvantages List three:
(1) ____
(2) ____
(3) ____

45. Identify two states where large wind farms are producing power for generation of electricity. ____

46. **(True/False)** In many locations wind provides an alternative to the use of nuclear or coal-fired power plants.

47. **(True/False)** By the year 2000, wind farms will produce more electricity than hydroelectric power plants.

48. Describe one way that the wind energy could be stored so that energy is available when the wind doesn't blow. ____
____
____

Biomass [479]

49. Any matter which is produced by living organisms and used as a source of energy is ______________. Give four examples. (1) ___________ (2) ___________
(3) ______________________________ (4) ___________________________

50. According to the U.S. Forest Service, forests can replace **(1/2, 1/3, 1/4)** of the oil imported each year. In the 1980s wood became an important source of energy for ____________________________________.

Name ______________________ **Date** ________ **Class Period** ____

51. The major source of biomass in Hawaii is ______________________ which is a waste product of the ______________ industry.

52. **Bagasse** pellets have the same energy value as ______________.

53. **Cogeneration** refers to the use of one source of energy for the production of __________________ and ______________.

54. While conventional power plants are only ________ percent efficient, cogeneration systems increase the efficiency to _________ percent.

55. A few years ago eastern Colorado farmers growing ______________ burned the excess straw. Now they sell it to a company that makes __________________.

56. Compare "strawloggs" to coal in the following ways.
 A. heat content: ____
 B. pollution produced during combustion: _____________________________
 C. use of ashes: ____

57. Biomass can be converted to ________________ which can be used to make gasohol. The major crop used to produce alcohol is ___________. The production of alcohol from farm crops **(will/will not)** provide enough fuel for vehicles to make the U.S. energy independent.

The Economics of Alternative Energy Sources [480]

58. Explain why more people don't use electricity produced by renewable energy sources. ____
____

59. The Department of Energy's goal is _______ percent of energy supplied by renewable resources by the year 2000.

60. Imagine that you are the next President of the United States. What steps would you take to insure an adequate supply of energy to meet our future needs?
____
____
____
____

Name ______________________________ Date ________ Class Period ____

Unit V — 1.4 Brunner Island and Other Solutions to Thermal Pollution

Introduction [484]

1. Combustion in a fossil-fueled power plant and fission in a nuclear power plant release energy that heats __________. The steam turns the blades of a __________ which is connected to a __________________. The rotation of a coil of copper wire causes _______________ to flow along the wire. This flow of electrons is called _________________.

2. Identify the correct amount of energy that:
A. The amount of energy converted to electricity in a nuclear power plant is _____ percent.
B. The amount of energy converted to electricity in a modern fossil-fueled power plant is _______ percent.
C. The amount of unused energy remaining in steam after it has passed through the turbine is ________ percent.
D. The amount of energy that is lost to the atmosphere as heat is _______ percent.

Efficiency of a Fossil-Fueled Steam-Electric Power Plant [484]

3. Match the following descriptions with the proper term.

___ A. a coil of wire in a magnetic field.	**(1) condenser**
___ B. changes steam to water.	**(2) electricity**
___ C. a large wheel with blades.	**(3) generator**
___ D. the flow of electrons.	**(4) turbine**
___ E. structure turned by the force of steam.	

Once-Through-Cooling Method [485]

4. Water for the once-through-cooling method comes from a __________, ________, or __________.

5. After the water is used in the once-through-cooling method, it is sent ___________ _________________________.

6. The once-through-cooling method is **(less/more)** expensive than other cooling methods.

7. Because of the increased temperature of the water, the level of dissolved oxygen **(decreases/increases)**. The warmer water causes the fish's metabolism rate and need for oxygen **(decreases/increases)**.

8. The warmer temperatures may cause a decline in the population of some species of algae and plants that are important ____________________.

9. The increased temperature may make some species of fish less resistant to ____________ or more sensitive to ____________________ or ______________.

10. This action between heated water and disease or heated water and chemicals is known as a ________________ effect.

Name ______________________ **Date** ________ **Class Period** ____

11. If the power plant must shut down for repairs, the change in ________________ might cause a fish kill.

12. Power plants and industries are not allowed to raise the natural temperature of water in a river or a stream more than ____°**F** and the water must never be above ____°**F**.

13. Cooling water can only be returned to the river if ______________________________

Cooling Towers [485]

14. More than _____ percent of water used by industries is used for cooling.

15. To prevent thermal pollution, some industries and power plants build ________________.

16. Indicate whether each process or structure is associated with a **wet tower W** or a **dry tower D**.
    _____ A. water flows through pipes.
    _____ B. water falls over a series of baffles.
    _____ C. requires a fan for cooling.
    _____ D. evaporation is a major problem.
    _____ E. is more expensive to build and operate.

Cooling Ponds or Reservoirs [485]

17. Cooling ponds, such as the one built by Kansas City Power and Light, are also used for ________________.

Waste Heat Greenhouses [486]

18. The warm water from power plants is piped through ________________ systems in greenhouses.

19. Give one disadvantage and one advantage of using waste-heat from power plants to heat greenhouses.
    Disadvantage: ______________________________
    Advantage: ______________________________

Aquaculture with Waste Water (447)

20. At Brunner Island Pennsylvania Power and Light combines fish farming or ________________ with power production.

21. The 80°F water at Brunner Island is ideal for raising ________________. In Canada, water from a power plant is used to raise ________________.

22. List two advantages of growing fish in raceways where the temperature and diet of the fish can be controlled.
    (1) ______________________________
    (2) ______________________________

Name ____ Date __________ Class Period ____

Unit V — 1.5 Three-Mile-Island

[page 488]

1. A study of cancer rates was not made until _____ years after the accident at Three Mile Island Nuclear Power Plant.

2. The type of cancer might be expected to increase five years after exposure to radiation? ______________ Other cancers may take ____to ____ years to develop.

3. Researchers at Columbia University released results of a study in 1990. Did they find an increase in leukemia? ____ Did they find an increase in cancer in children? _____

4. The best method of study would have been a __________ and __________ study. Why wasn't this method used? ____ ____

5. A major problem that affected the results of a scientific study to determine the effects of the accident was the _______________ of the human population.

6. The method used by the Pennsylvania Department of Health compared the number of cancer cases **observed** in the population to the number of cancer cases that were _______________.

7. Cancer registry files were not begun until 19____. Other sources of information used by the Department of Health included , ___________ data, ____________ records, ___________ certificates, and surveys of ___________________.

8. Complete the following chart:

Cancer Deaths	Observed	Expected
"Downwind Communities"		
10–mile radius of TMI		
20–mile radius of TMI		

9. In which region was the number of cancer deaths observed greater than the number expected? _____________________________

10. Considering the total population of the "downwind" communities was 25,000, is it likely that the difference in number of cancer deaths is due to the accident at Three Mile Island? ____

11. Do you think that the differences in the number of cancer deaths within a 10 and 20 mile radius of TMI can be due to the radiation released during the accident? ____

Name ______________________________ **Date** ________ **Class Period** ____

12. Complete the following chart:

New Cases	Observed	Expected
"Downwind Communities"		
Leukemia		
Pregnant women		
Babies		

13. Is the difference between the observed and expected new cases of cancer (12 additional cases) a "statistically significant" difference? _____ Do scientists think that the accident at Three-Mile-Island is responsible for any of the differences shown above? ______________________________

14. Give two reasons why the additional ten cases of cancer found in Newberry Township are not thought to be related to the accident at TMI.
(1) ______________________________
(2) ______________________________

15. During their lifetime _____ out of _____ people will develop cancer.

16. Of the nine leukemia deaths reported after the TMI accident, at least three probably were not caused by the radiation released at the time of the accident. Why?
____
____

17. The study conducted by the Pennsylvania Department of Health and the study known as the Aamodt survey led to different conclusions about the increase in the rates of cancer. Identify the study that is described by each of the following statements.

(1) Aamodt Survey **(2) State Department. of Health**
(3) both studies **(4) neither study**

	A.	a door to door survey.
	B.	concluded that the cancer death rate was several times higher than normal.
	C.	did not consider age and sex of the population studied.
	D.	included cancer deaths although the cancer had been diagnosed before the TMI accident.
	E.	included cancers that occurred in women who were pregnant at the time of the TMI accident or became pregnant shortly after the accident.
	F.	considered all cancers even those that are not thought to be caused by radiation.
	G.	was requested by the Advisory Panel for Decontamination of TMI Unit 2.
	H.	a study done by a group of local residents
	I.	study that is considered valid by most scientists.
	J.	data collected was biased.
	K.	did not include former residents and did not exclude newcomers
	L.	used the 1980 population of the TMI area.

Name ______________________________ **Date** ________ **Class Period** ____

18. Scientists expected the number of cancer cases to increase in the area of TMI, with or without an accident because:
(1) the population has ________________ since the accident.
(2) the number of cancer cases has __________________ in many regions.
(3) methods of ____________ and (4) ____________ cancer have improved.
(5) people are ____________ longer.

19. Do you think that a person who was a long-term heavy smoker and died of lung cancer should have been included in the study? _____ Why or why not?
____
____

20. Is there evidence that the TMI accident caused an increase in the cancer death rate?
____
____

21. Why are more studies needed? ______________________________
____
____

Introduction [493]

1. Identify the country that uses more energy per person than any other country. __________________________.

2. A country's standard of living depends upon its ability to ________________ ________________ which often requires huge supplies of ______________.

3. Oil and natural gas are sources of ____________ and ____________________.

4. List 6 types of products that are made from petrochemicals.
(1) ________________ (2) ________________ (3) ________________
(4) ________________ (5) ________________ (6) ________________

5. How is the American lifestyle of the 1970s and 1980s different from the lifestyle of the 1940s and 1950s? ____

6. What factors do consumers consider when purchasing a product?

What factors should be considered? ____________________________________

The Housing Project [494]

7. Identify the source of each of the following building materials:
Aluminum: ____
PVC (polyvinylchloride): ____________________________________

8. Identify factors that must be considered when comparing the environmental impact of using a specific type of construction material.

9. Based on the information in the chart on page 500, which material in each of the following pairs requires the greatest amount of energy to produce?
( ) **ABS plastic or** ( ) **aluminum**
( ) **HDPE plastic or** ( ) **steel**

10. Based on the information in the chart on page 500, which material in each of the following pairs produces the greatest amount of air and water pollutants?
( ) **ABS plastic or** ( ) **aluminum**
( ) **HDPE plastic or** ( ) **steel**

11. Compare the environmental impacts caused by making a canoe of ABS plastic with the impacts caused by making the canoe of aluminum. If your only concern in the purchase of a canoe was the environmental impact caused by its manufacture, which canoe would you buy? ______________________________

12. Why do builders often prefer plastic products offer over other building materials?
(1) ______________________________
(2) ______________________________

13. Why do homeowners often prefer patio furniture made of plastic rather than metal or wood?
(1) ______________________________
(2) ______________________________
(3) ______________________________

The Tiger in Your Tank [495]

14. What characteristics of plastics makes them desirable for use in the manufacture of a car? ______________________________

15. Plastics are made from crude oil. How has the use of plastics by the auto industry helped reduce the amount of crude oil that the United States needs to import?
____
____

The Shirt in Your Closet [496]

16. Which of the following fibers is made from a nonrenewable resource?
( ) **cotton** ( ) **polyester** ( ) **wool**

17. The cotton farmer depends upon fossil fuels. ____________ is used to produce the steel used in making the farm machinery. The machines will not run without fuel made from ____________. Growing cotton requires ____________ that are made from natural gas and ____________ that are made from petrochemicals.

18. It requires more energy to manufacture a shirt that is made of:
( ) **a cotton/polyester blend** ( ) **100% cotton**

19. It requires more energy to maintain a shirt that is made of:
( ) **a cotton/polyester blend** ( ) **100% cotton**

To Market, To Market [497]

20. Compare the environmental impacts caused by making a plastic cup with the impacts caused by making a cup of paper. If your only concern in the purchase of the cups was the environmental impact caused by its manufacture, which would you choose — paper or plastic? ____________ Justify your choice.
____
____
____

21. Which milk carton has the least environmental impact? ____________________

22. Which milk carton would you choose? ____________________
What factors influenced your decision? ____________________
____

23. Given a choice of a paper or polystyrene meat tray, which would you choose? ____________________ What factors influenced your decision? ____________________
____
____

24. Which type of beverage container has the least environmental impact? ____________________ Which type of beverage container would you choose? ____________________ What factors influenced your decision? ____________________
____
____

25. Which material requires the most energy to produce from virgin (raw) materials — aluminum, glass or plastic? ____________________

26. Which material saves the most energy when recycled —aluminum, glass or plastic? ____________________

27. Which grocery bag has a lower environmental impact? ____________________ Which bag did you choose? ____________________ Justify your choice. ____________________
____
____

28. What changes could or should you make to reduce your impact on planet Earth?
____
____
____
____
____

Burn It or Bury It? That is the Question! [502]

1. The average U.S. citizen throws away ____ pounds of solid waste each day. If waste from construction sites is included, the waste produced per person is ____ pounds per day.

2. Assume that the members of your family are average. How much waste is produced by your family during one year? ________________________

3. Waste produced in most cities is: ___ burned ___ buried.

4. According to the National Solid Waste Management Association, how many of our landfills will still be able to accept trash ten years from now? ________________

   ____

5. Identify two problems associated with the use of sanitary landfills.
   (1) ____
   (2) ____

6. List the EPA requirements that landfills must meet in order to accept trash today.
   (1) ____
   (2) ____
   (3) ____
   (4) ____

7. Most of the waste going to landfills is ______________.

8. Give two advantages of burning instead of burying as a means of waste disposal.
   (1) ____
   (2) ____

9. Compare the energy content of the following wastes. Circle the waste in each pair that has the most energy in Btus/pound.
   (food waste/grass) **(plastic/wood)** **(rubber/wood)**

10. Compare the amount of domestic waste burned in the United States to the amount burned at waste-to-energy plants in Denmark, Sweden, Switzerland and Japan.

    ____
    ____

11. The energy produced at waste-to-energy plants may be used to produce ______________ or ______________ that is used by industries.

12. The EPA recommends that local governments burn _____ percent of their solid waste.

13. Identify two concerns associated with waste-to-energy plants.
    (1) ____
    (2) ____

Resource Recovery [503]

14. ______________________ is the salvaging of usable materials from solid waste. ______________ is the use of "waste" materials that have been recovered from the solid waste stream.

15. In 1991, _____ percent of the solid waste produced in the United States was recycled.

16. Before 1975 most soft drinks were packaged in _____ bottles that were returnable. Today most soft drinks are packaged in _____________ or _____________.

17. The advantages of recovering and recycling aluminum and other materials include: (1) saving ________________________ and _____________. (3) reducing __________________________.

18. Recycling an aluminum can requires only ____ percent of the energy required to make an aluminum can from bauxite and reduces emissions of _____________ ______________ According to ALCOA, recycling aluminum creates ________.

19. According to a study, the bottle deposit required by the state of New York has: (1) reduced the ____ (2) provided nearly 4000 _____________.

20. How many states have "bottle bills"? _______ Is there a deposit law in your state? _________

21. Identify three types of materials in addition to aluminum that may be recycled. (1) _________________ (2) _____________ (3) ______________________

22. High grade office paper should be **(recycled/burned)**.

23. Explain how resource recovery and recycling affect an ecosystem.
____
____
____
____

24. Why have some states banned the sale of juice boxes? _______________________
____

25. Does your state have a mandatory recycling law? _______
If so, what type of items are recycled? __________________________________
Does your family recycle? _______
What happens to the waste that is not recycled? ___________________________
Do you think that recycling should be voluntary or required? ________________

26. Are leaves and grass banned from landfills in your state? _________
Leaves and grass can be recycled into ________________.

27. Describe the "pay-as-you-throw" policy and explain how it affects the disposal of solid waste. ____
____
____

28. Explain why is it essential that the phrase "reduce, reuse, recycle" must become our lifestyle. ____
____
____

Reduce and Reuse [505]

29. People in North America produce __________ the amount of waste produced by other industrialized countries.

30. The greater amount of solid waste is produced by cooks making breakfast in kitchens **(in the city/on the farm)**.

31. In many cases the greatest amount of energy is used to **(package/produce)** convenience foods.

32. About ________ percent of the volume of domestic waste is packaging materials.

33. What is the EPA's top priority in solid waste management? ____________________

34. How did McDonald's Corporation reduce solid waste? ______________________
____ Is this source reduction? _____

35. Which is the best way to avoid solid waste — reuse or recycle? ______________
Explain why. ____
____
____